THE QUANTUM THEORY

THE
QUANTUM THEORY

BY

FRITZ REICHE

PROFESSOR OF PHYSICS IN THE UNIVERSITY OF BRESLAU

TRANSLATED BY H. S. HATFIELD, B.Sc., Ph.D., AND
HENRY L. BROSE, M.A.

WITH FIFTEEN DIAGRAMS

METHUEN & CO. LTD.
36 ESSEX STREET W.C.
LONDON

First Published in 1922

CONTENTS

v

THE QUANTUM THEORY

INTRODUCTION

THE old saying that small causes give rise to great effects has been confirmed more than once in the history of physics. For, very frequently, inconspicuous differences between theory and experiment (which did not, however, escape the vigilant eye of the investigator) have become starting-points of new and important researches.

Out of the well-known *Michelson-Morley* experiment, which, in spite of the application of the most powerful methods of exact optical measurement, failed to show an influence of the earth's movement on the propagation of light as was predicted by classical theory, there arose the great structure of *Einstein's* Theory of Relativity. In the same way the trifling difference between the measured and calculated values of black-body radiation gave rise to the Quantum Theory which, formulated by *Max Planck*, was destined to revolutionise in the course of time almost all departments of physics.

The quantum theory is yet comparatively young. It is therefore not surprising that we are confronted with an unfinished theory still in process of development which, changing constantly in many directions, must often destroy what it has built up a short time before. But under such circumstances as these, in which the theory is continually deriving new nourishment from a fresh stream of ideas and suggestions, there is a peculiar fascination in attempting to review the life-history of the quantum theory to the present time and in disclosing the kernel which will certainly outlast changes of form.

1

CHAPTER I

The Origin of the Quantum Hypothesis

§ 1. Black-Body Radiation and its Realisation in Practice

THE Quantum Theory first saw light in 1900. When, in the years immediately preceding (1897-1899), *Lummer* and *Pringsheim* made their fundamental measurements [1] of black-body radiation at the *Reichsanstalt*, they could have had no premonition that their careful experiments would become the starting-point of a revolution such as has seldom occurred in physics.

In the field of heat radiation chief interest at that time was centred in the radiation of a black body (briefly called "black-body" radiation), that is, of a body which absorbs completely all radiation which falls on it and which thus reflects, transmits, and scatters [2] none. We may shortly call to mind the following facts. It is known that any body at a given temperature sends out energy in the form of radiation into the surrounding space. This radiation is not energy in a single simple form but is made up of a number of single radiations of different colours, i.e. of different wave-lengths λ or of different frequencies [3] ν. In other words, it forms in general a spectrum in which radiations of all frequencies between $\nu = 0$ and $\nu = \infty$ are represented. Further, these radiations are present in varying "intensities." We define this term thus. Consider the radiation emitted from unit surface of the body per second in a certain direction; break it up spectrally and cut out of the spectrum a small frequency interval $d\nu$ such that it contains all frequencies between ν and $\nu + d\nu$. The energy of radiation E_ν thus sliced out (namely, *the emissivity of the body for the frequency ν*) may be defined in the following terms : [4]

$$E_\nu = 2\pi \mathrm{K}_\nu d\nu \qquad . \qquad . \qquad . \quad (1)$$

2

provided that—as we shall assume for the sake of simplicity
—the surface of the body emits uniform and unpolarised
radiation in all directions.

The magnitude K_ν thus defined is called the intensity of
radiation of the body for the frequency ν. It is in general a
more or less complicated function of the frequency ν, of the
absolute temperature of the body T, and of the inherent
properties of the body. The black body alone is unique in
this respect. For its radiation and therefore its K_ν is, as
G. Kirchhoff[5] was the first to point out, dependent only
on the frequency ν and the absolute temperature T, that is,
mathematically,

$$K_\nu = f(\nu, T) \qquad . \qquad . \qquad . \qquad . \qquad (2)$$

This formula which gives the relation between the intensity
of radiation from a black body, the temperature, and the
"colour" is called the radiation formula or the law of radiation of a black body.

To calculate this relationship on the one hand and to
measure it on the other were unsolved problems at that
time. Unimpeachable measurements were of course possible
only if one could succeed in constructing a black body which
approached sufficiently near the theoretical ideal. This important step, the realisation of the black body, was taken by
O. Lummer and W. Wien[6] on the basis of Kirchhoff's[7]
Law of Cavity Radiation, which states : *In an enclosure
or a cavity which is enclosed on all sides by reflecting walls,
externally protected from exchanging heat with its surroundings,
and evacuated, the condition of " black radiation" is automatically set up if all the emitting and absorbing bodies at the
walls or in the enclosure are at the same temperature.* In a
space, therefore, which is hermetically surrounded by bodies
at the same temperature T and which is prevented from exchanging heat with its surroundings, every beam of radiation
is identical in quality and intensity with that which would be
emitted by a black body at the temperature T.

Lummer and *Wien*, therefore, had only to construct a
uniformly heated enclosure with blackened walls having a
small opening. The radiation emitted from this opening was
then "black" to an approximation which was the closer

the smaller the opening, that is, the less the completeness of the enclosure was disturbed. The manner in which the intensity K_ν of the black radiation thus realised depended on the frequency ν and the temperature T had next to be determined. The above-mentioned investigation of *Lummer* and *Pringsheim* was devoted to this purpose.

§ 2. The Stefan-Boltzmann Law of Radiation and Wien's Displacement Law

While experimental research was proceeding on its way, theory was not idle, for valuable pioneer work was being done inasmuch as two fundamental laws were set up. In the first place, *L. Boltzmann*[8] proved, with the help of thermodynamics, the law previously enunciated by *Stefan*,[9] that the sum-total of the radiation from a black body, taking all the frequencies together, namely, the quantity $K = \int_0^\infty K_\nu d\nu$, is proportional to the fourth power of its absolute temperature : [10]

$$K = \gamma . T^4 \ (\gamma = \text{const.}) \quad . \quad . \quad . \quad (3)$$

The laws proposed by *Wien*[11] entered more deeply into the question. *Wien* imagines the black radiation enclosed in a closed space with a perfectly reflecting piston as one wall, and then supposes the radiation to be compressed adiabatically, as in the case of gases (that is, no passage of heat to or from the cavity is allowed during the process), by infinitely slow movements of the piston. Now, if we express the change which this process causes in the energy of a definite colour interval $d\nu$ in two ways, and if we take into consideration that the waves reflected at the piston undergo a change of colour according to *Doppler's* principle, we succeed in limiting very considerably the unknown functional dependence of the quantity K_ν on ν and T. There is thus obtained a relation of the form [12]

$$K_\nu = \frac{\nu^3}{c^2} \ F\left(\frac{\nu}{T}\right) \quad . \quad . \quad . \quad (4)$$

in which c is the velocity of light *in vacuo*, the function F being left undetermined. From this, *Wien's* Displacement Law, the conclusion [13] may be drawn that the frequency

ν_{max} for which $\mathbf{K}_\nu$ (plotted as a function of ν) is a maximum is displaced towards higher values proportional to T as the temperature increases :

$$\nu_{\text{max}} = \text{const.} \, . \, T \qquad . \qquad . \qquad . \quad (4a)$$

If, as is usual in physical measurement, we use the wave-length $\lambda = \dfrac{c}{\nu}$ instead of the frequency as the variable, *Wien's* Law assumes a somewhat different form. For if we consider the radiant energy of a narrow range of wave-length $d\lambda$ corresponding to the frequency range $d\nu$, and write it in the form $E_\lambda d\lambda$, then $E_\lambda d\lambda = \mathbf{K}_\nu d\nu$, that is : $E_\lambda = \mathbf{K}_\nu \, . \, \dfrac{c}{\lambda^2}$. In place of (4) and (4a) we then get the relations :

$$E_\lambda = \frac{c^2}{\lambda^5} \, F\!\left(\frac{c}{\lambda T}\right) \qquad . \qquad . \qquad . \quad (5)$$

$$\lambda_{\text{max}} \, . \, T \, . \, \text{const.} = \delta \qquad . \qquad . \qquad . \quad (5a)$$

§ 3. Wien's Law of Radiation

To formulate the law of radiation it was therefore necessary only to evaluate the unknown function F in (4) or (5). But this was just the central point of the whole question, and the most difficult part of the problem.

Here, too, *Wien* made the first successful attack. On the basis of not entirely unobjectionable calculations, which were founded on *Maxwell's* law of distribution of velocities among gas molecules, he arrived at the following specialised form [14] of the function F :—

$$F = a \, . \, e^{-\beta \frac{\nu}{T}} \qquad (a \text{ and } \beta \text{ are two constants}).$$

Thus the law of radiation (4) assumes the form

$$\mathbf{K}_\nu = a \, \frac{\nu^3}{c^2} \, . \, e^{-\beta \frac{\nu}{T}} \qquad . \qquad . \qquad . \quad (6)$$

which is called *Wien's* Law of Radiation.

How far did experiment confirm these theoretical results ? While the *Stefan-Boltzmann* Law and *Wien's* Displacement Law were confirmed to a large extent by the observations of *Lummer* and *Pringsheim*,[15] both experimenters found *Wien's*

Law of Radiation confirmed only for high frequencies, that is, for short wave-lengths (more precisely, for large values of $\beta\frac{\nu}{T}$), and detected, on the other hand, *systematic discrepancies for small frequencies, that is, for long wave-lengths.*[16]   They maintained with unswerving persistence that these discrepancies were real in spite of objections from authoritative quarters. For while *F. Paschen* [17] imagined that he had proved by his work that *Wien's* Law of Radiation was universally valid, *Max Planck*, in his detailed theory of irreversible processes of radiation,[18] had arrived again at *Wien's* radiation formula by a more rigorous method.   Starting from *Kirchhoff's* Law of Cavity Radiation, according to which the presence of any emitting or radiating substance whatsoever in a uniformly heated enclosure produces and ensures the maintenance of the condition of black-body radiation, *Planck* chose as the simplest schematic model of such a substance a system of linear electromagnetic oscillators, and investigated the equilibrium of the radiation set up between them and the radiation of the enclosure.   This is to be understood as follows: Each of the *Planck* oscillators—as such we may, for example, assume bound electrons capable of vibration—possesses a fixed natural frequency ν and responds, on account of its weak damping, only to those waves of the radiation in the enclosure whose frequencies lie in the immediate neighbourhood of ν, while all other waves pass over it without effect.   The oscillator thus acts selectively, as a resonator, in just the same way as a tuning-fork of definite pitch commences to sound only when its own " proper " tone, or one very near it, is contained in the volume of sound which strikes it.   In this process of resonance, however, the oscillator exchanges energy with the radiation inasmuch as, on the one hand, it acts as a resonator in abstracting energy from the external radiation, and, on the other, it acts as an oscillator and radiates energy by its own vibration.   Hence a dynamic equilibrium is set up between the oscillator and the radiation of the enclosure, and, indeed, between just those waves of the radiation which have the frequency ν.   In this state of equilibrium the radiation of frequency ν acquires an intensity K_ν which, according to *Kirchhoff's* Law, *is equal to the intensity of black-body radiation at this temperature.* Secondly, the energy U of the oscillator passes in the course of time through all possible values, the mean value [19] $\overline{U}$ of which is found to be proportional to the intensity K_ν, a result which seems immediately plausible since the excitation of the oscillator will be greater the more intense the radiation that falls on it. The exact calculation of this relationship between $\overline{K}_\nu$ and $\overline{U}$ on the basis of classical electrodynamics—this is the first part of *Planck's* calculations—leads to the fundamental formula :

$$K_\nu = \frac{\nu^2}{c^2} \cdot \overline{U} \qquad . \qquad . \qquad . \qquad (7)$$

In the second part *Planck* [20] determined $\overline{U}$, although by a method that is not free from ambiguity, as a function of ν and T on the basis of the second law of thermodynamics. He obtained

$$\overline{U} = a\nu e^{-\beta \frac{\nu}{T}} \qquad . \qquad . \qquad . \qquad (8)$$

The combination of (7) and (8) gives us *Wien's* Law of Radiation (6).

§ 4. The Quantum Hypothesis. Planck's Law of Radiation

Lummer and *Pringsheim*, however, refused to surrender. In a fresh investigation [21] in 1900 they showed that in the region of long waves *Wien's* radiation formula undoubtedly did *not* agree with the results of observation. As a result of this, *Planck*, in an important paper [22] which must be regarded as marking the creation of the quantum theory, decided to modify his method of deducing the law of radiation, namely, by altering the expression (8) which gives the mean energy of the oscillator, but which is not unique. He proceeded as follows.[23] In order to distribute the whole available energy among the oscillators, he imagined this energy divided into a discrete number of finite " elements of energy " (*energy quanta*) of magnitude ϵ, and supposed these energy quanta to be distributed at random among the individual oscillators exactly as a given number of balls, say 5, may be distributed among a certain number of boxes, say 3. Each such distribution (of 5 balls among 3 boxes) may obviously be carried out in a number of different ways, whereby, however, we are not concerned with which *particular* balls lie in which *particular* boxes, but with the *number* contained in each.[24]   Now since each such " distribution " corresponds to a definite state of the system, it follows from what has just been said that each condition may be realised in a number of different ways, that is, each condition is characterised by a certain number of possibilities of realisation.   This number is called by *Planck* the thermodynamic probability W of the condition in question.   For it is obvious that the probability of a condition or state is the greater, i.e. it will occur the more frequently, the greater the number of ways in which it may be realised.   By means of the usual formulæ of permutations and combinations, of which the latter alone come into consideration here, it was possible to calculate the probability of any given distribution of the elements of energy among the oscillators, and thus also the probability of a given energetic condition of the system of oscillators as a function of the mean energy $\overline{U}$ of an oscillator and of the energy quantum.   Now, *L. Boltzmann*[25] has given an extremely fertile rule, which connects the probability of state W of a system with its *entropy* S, a magnitude which, as is well known, plays a similar rôle in the second law of thermodynamics to that played by energy in the first.   Thus S was obtained as a function of $\overline{U}$ and ϵ.   If now, on the other hand, one applied the second law itself, which expresses the entropy S as a function of the mean energy $\overline{U}$ and the absolute temperature T, the following result was obtained by this circuitous process : the entropy, as an auxiliary magnitude, was eliminated, and a *relation between* $\overline{U}$, T, *and* ϵ was gained. This fundamental result, first obtained by *Planck*, is as follows :—

$$\overline{U} = \frac{\epsilon}{e^{\frac{\epsilon}{kT}} - 1} \qquad (k \text{ being a constant}) . \quad . \quad (9)$$

But from (7) and *Wien's* Displacement Law (4) it follows that for the mean energy $\overline{U}$ of an oscillator, a relationship of the following form exists :—

$$\overline{U} = \nu . F\left(\frac{\nu}{T}\right) . \qquad . \qquad . \qquad . \quad (10)$$

A comparison of (9) and (10) shows that $\overline{U}$ assumes the form required by (10) only when ϵ is set proportional to ν, the frequency. This is an essential point of *Planck's* Theory: if *we are to remain in agreement with* Wien's *Displacement Law, the energy element ϵ must be set equal to* $h\nu$

$$\epsilon = h\nu \qquad . \qquad . \qquad . \qquad . \qquad (11)$$

The constant h, which, on account of its dimensions (energy × time), is called *Planck's Quantum of Action*, has played, as we shall see, a rôle of undreamed-of importance in the further development of the quantum theory.

By combining the formulæ (7), (9), and (11) the renowned radiation law of *Planck* follows at once :—

$$K_\nu = \frac{h\nu^3}{c^2} \cdot \frac{1}{e^{\frac{h\nu}{kT}} - 1} \qquad . \qquad . \qquad . \qquad (12)$$

which *Planck* first deduced in the year 1900 in the manner above described, that is, by the hypothesis of energy quanta. In the same year as well as in the following year this Law of Radiation was confirmed very satisfactorily by *H. Rubens* and *F. Kurlbaum* [26] for long waves, and by *F. Paschen* [27] for short waves. The later measurements of radiation emitted by black bodies,[28] particularly the exact work carried out by *E. Warburg* and his collaborators at the Reichsanstalt, have also demonstrated the validity of *Planck's* formula. In opposition to this, *W. Nernst* and *Th. Wulf*,[29] as the result of a critical review of the whole experimental material available up to that date, have recently shown the existence of deviations (up to 7 per cent) between the measured and the calculated values according to *Planck's* formula, and hence feel themselves constrained to decide against the exact validity of *Planck's* formula. Whatever view is taken of this criticism, it is at any rate a powerful incentive to take up anew the measurement of the radiation emitted by black bodies with all the finesse and precautions of modern experimental science, and thereby to decide finally the important question whether *Planck's* Law is exactly valid or not.

For short wave-lengths, i.e. high frequencies (more exactly, for high values of $\frac{h\nu}{kT}$), *Planck's* formula assumes the form

$$K_\nu = \frac{h\nu^3}{c^2} \cdot e^{-\frac{h\nu}{kT}} \quad . \quad . \quad . \quad (13)$$

and thus passes over into *Wien's* Law (cf. formula (6), which, as we have seen, was confirmed by experiment for these frequencies). In the other limiting case, i.e. for long waves, low frequencies (more exactly for small values of $\frac{h\nu}{kT}$), *Planck's* formula assumes the form

$$K_\nu = \frac{\nu^2}{c^2} kT \quad . \quad . \quad . \quad (14)$$

as is easily found by developing the exponential function $e^{\frac{h\nu}{Tk}}$ as a series. This limiting law, which has been confirmed in the region of long wave-lengths, had been given previously by *Lord Rayleigh*.[30] *Planck's* formula thus contains *Wien's* Law and *Rayleigh's* Law as limiting cases.

If we use the wave-length λ instead of the frequency ν, *Planck's* Law takes the form

$$E_\lambda = \frac{hc^2}{\lambda^5} \cdot \frac{1}{e^{\frac{hc}{k\lambda T}} - 1} \quad . \quad . \quad . \quad (15)$$

To make this clear, the intensity of radiation E_λ is plotted in Fig. 1 as a function of λ for various values of T. The curves which exhibit K_ν as a function of ν have a quite similar appearance. The maximum of the E_λ-curves lies at the point at which $\frac{hc}{k\lambda T}$ has the value $4·9651$.

It follows that

$$\lambda_{max} \cdot T = \frac{ch}{4·9651 \cdot k} = 6·042 \times 10^9 \cdot \frac{h}{k} = \delta \quad . \quad (16)$$

a relation, which is identical in form with *Wien's* Displacement Law (5a).

For the total radiation we get from (12) or (15)

$$K = 2\int_0^\infty K_\nu d\nu = 2\int_0^\infty E_\lambda d\lambda = \frac{2\pi^4 k^4}{15c^2 h^3} \cdot T^4 = \gamma \cdot T^4 \quad . \quad (17)$$

an equation which gives expression to the *Stefan-Boltzmann Law* [31] (3).

From (16) and (17) we recognise that the measurement (a) of the total radiation (K) and (b) of the wave-length of the maximum (λ_{max}), at a fixed known temperature, allows us to calculate the two constants h and k of the radiation formula.[32] From *Kurlbaum's* measurements of the *Stefan-Boltzmann constant* γ, which were available at that time, and from the constant δ of *Wien's* Displacement Law (measured by *Lummer* and *Pringsheim*) *Planck* [33] found the following values :

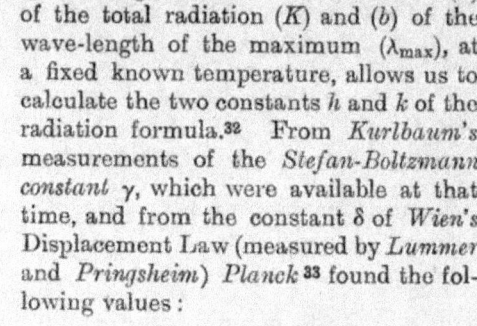

$$h = 6 \cdot 548 \times 10^{-27} [\text{erg . sec.}]$$
$$k = 1 \cdot 346 \cdot 10^{-16} \left[\frac{\text{erg}}{\text{deg.}} \right] \quad . \quad (18)$$

Corresponding to the varying values which have been found in the course of time for the constants γ and δ, the values of h and k have undergone changes which are not worth while recording here. For particularly the measurement of the total radiation —as we see from

FIG. 1.

the strongly varying values given in note 15—has not yet reached a sufficient degree of certainty, to allow a very accurate calculation of the two radiation constants h and k to be based on the *Stefan-Boltzmann* constant. Methods which allow h to be determined with undoubtedly much greater accuracy will be described later.

§ 5. Consequences of Planck's Theory

The deduction of the radiation formula and the determination of its constants did not, however, exhaust the successes of *Planck's* new theory; on the contrary, important relationships of this theory to other departments of physics became immediately revealed. For it was found [34] that the constant k of the radiation formula is nothing other than the quotient of the absolute gas constant R (which appears in the equation of state of an ideal gas) and the so-called *Avogadro* number N, i.e. the number of molecules in a grammolecule.

$$k = \frac{R}{N} \qquad . \qquad . \qquad . \qquad . \qquad (19)$$

As the value of R is sufficiently accurately known from thermodynamics

$$\left(R = 8\cdot31 \times 10^7 \left[\frac{\text{erg}}{\text{deg.}}\right] = 1\cdot98 \left[\frac{\text{cal.}}{\text{deg.}}\right] \right),$$

Planck,[35] by making use of the radiation measurements, was able to calculate the value of N. By using (18) he found

$$N = 6\cdot175 \times 10^{23} \qquad . \qquad . \qquad . \qquad (20)$$

The agreement of this value with the values deduced by quite different methods is very striking.[36] *Avogadro's* Law forms the bridge to the electron theory. For it is known that the electric charge which travels in electrolysis with 1 gramme-ion, that is, with N-ions, is a fundamental constant of nature, which is called the *Faraday*. Its value was, according to the position of measurements at that time, $9658 . 3 . 10^{10}$ electrostatic units (the value nowadays accepted [37] is $9649\cdot4 . 2\cdot999 . 10^{10}$). If now each monovalent-ion carries the charge e of the electron, the equation

$$Ne = 9658 . 3 . 10^{10} \qquad . \qquad . \qquad . \qquad (21)$$

must hold. From this, by using (20), we get

$$e = 4\cdot69 \times 10^{-10} \quad \text{electrostatic units} \qquad . \qquad (22)$$

The value of the electron charge thus calculated by *Planck* from the theory of radiation differs only by about 2 per cent from the latest and most exact measurements of *R. A. Millikan*,[38] who found the value

$$e = 4\cdot774 . 10^{-10} \quad \text{electrostatic units.} \qquad . \qquad (23)$$

A truly astonishing result.

CHAPTER II

The Failure of Classical Statistics

§ 1. The Equipartition Law and Rayleigh's Law of Radiation

IF these great successes had justified faith in *Planck's* Theory, it was also soon recognised—as had already been emphasised by *Planck* in his first papers—that the central point of the theory lay in the Quantum Hypothesis, i.e. in the novel and repulsive conception, that the energy of the oscillators of natural period ν was not a continuously variable magnitude, but always an integral multiple of the element of energy, that is $\epsilon = h\nu$. The recognition of the necessity of this hypothesis has forced itself upon us more and more in the course of time, and has become established, more especially through indirect evidence, inasmuch as *every attempt to work with the classical theory has led logically to a false law of radiation.* For when *Planck* turned the radiation problem. into a problem of probability—for a definite amount of energy was to be divided among the oscillators according to chance, and the mean value $\bar{U}$ of the energy of an oscillator was to be calculated—it became possible to apply the methods of the statistical mechanics founded by *Clerk Maxwell, L. Boltzmann,* and *Willard Gibbs.* And the application of these methods to the case in question appeared to be demanded from the start, if the standpoint, self-evident in classical physics, that the energy of the oscillator could assume in continuous sequence all values between 0 and ∞ were adopted. What, then, did statistical mechanics require? One of its chief laws is the law of the equipartition of kinetic energy,[39] *according to which in a state of statistical equilibrium at absolute temperature T every degree of freedom of a mechanical system, however complicated, possesses the mean kinetic*

13

energy $\frac{1}{2}kT$. In this expression the constant k is defined by (19), and is thus the same constant as that which appears in the Law of Radiation. A system of f degrees of freedom, therefore, possesses at a temperature T a mean kinetic energy $f \cdot \frac{1}{2}kT$. For example, the atom of a monatomic gas is a configuration which possesses three degrees of freedom, if we regard it from the point of view of mechanics as a mass-point. Its kinetic energy at the temperature T has therefore a mean value [40] $\frac{3}{2}kT$, independent of its mass, a result which has been known in the kinetic theory of gases since the time of *Maxwell*, and which is deduced as a consequence of his law of distribution of velocities.

Planck's linear oscillator, which is essentially identical with an electron vibrating in a straight line, possesses one degree of freedom; its kinetic energy at the temperature T has therefore the mean value $\frac{1}{2}kT$. Now the mean potential energy of the oscillator is equal to its mean kinetic energy.[41] As a result, its mean total energy (kinetic plus potential) has the value

$$\bar{U} = kT \qquad . \qquad . \qquad . \qquad . \quad (24)$$

This result of classical statistics, when combined with the relation (7) deduced from classical electrodynamics, gives *Rayleigh's* Law of Radiation

$$\mathbf{K}_\nu = \frac{\nu^2}{c^2} kT \qquad . \qquad . \qquad . \quad (25)$$

which, as we saw (cf. (14)), is contained in *Planck's* Law of Radiation as a limiting case for small values of $\frac{h\nu}{kT}$, that is, for long waves or high temperatures.

This Law of Radiation of *Rayleigh* which, deduced as it is from the fundamental principles of classical statistics and electrodynamics, should be able to claim general validity for all frequencies and all temperatures, stands none the less in glaring contradiction to observation. For while all observed curves of distribution of energy of a black body (i.e. $\mathbf{K}_\nu$ plotted as a function of ν, T being constant) *always show a maximum*, the curve expressed by (25) rises without limit for rising values of ν, and therefore gives for the sum $K = 2\int_0^\infty \mathbf{K}_\nu d\nu$ an infinitely large value.

§ 2. Fruitless Attempts at Improvement

From very different quarters and in the most varied ways attempts were made, as time went on, to escape from *Rayleigh's* Law without discarding classical statistical mechanics. All in vain. Thus *J. H. Jeans*,[42] without making use of a "material" oscillator, considered only the radiation *as such* in an enclosure, and distributed the whole energy of radiation according to the Law of Equipartition over the individual "degrees of freedom of radiation" (which are here the individual vibrations that are possible in an enclosure). Further, *H. A. Lorentz* [43] deduced in a penetrating investigation the thermal radiation of the metals, starting from the conception that the free "conduction electrons," which carry the current, produce the radiation by their collisions with the atoms, and applying the Law of Equipartition to the motion of these electrons. The problem was attacked in a somewhat different fashion by *A. Einstein* and *L. Hopf*.[44] They imagined the *Planck* oscillator firmly attached to a molecule, and then considered this complex exposed to the radiation and the impacts of other molecules. The Law of Radiation could then be deduced from the condition that the impulse, which the impacts of the molecules give to the complex, must not on the average be changed by the impulses, which the radiation gives to the oscillator. We may also mention a paper of *A. D. Fokker* [45] which was supplemented by *M. Planck*.[46] In this, by the aid of a general law due to *Einstein*, the statistical equilibrium between the radiation and a large number of oscillators was examined on the basis of the classical theories. All these different ways ended, however, at the same point; they all led to *Rayleigh's* Law. And finally, at the Solvay Congress in Brussels in 1911, *H. A. Lorentz* [47] showed, in the most general manner imaginable, that we arrive of necessity at this wrong law, if we assume the validity of *Hamilton's* Principle and of the Principle of Equipartition for the totality of the phenomena (of mechanical and electromagnetic nature) which take place in an enclosure containing radiation, matter, and electrons. Only in the limiting case of high temperatures or small frequencies do the results of the classical theory agree with the results of observation.[48]

CHAPTER III

The Development and the Ramifications of the Quantum Theory

§ 1. The Absorption and Emission of Quanta

AS stated above, the conviction was bound to establish itself that every attempt to deduce the laws of radiation on the basis of classical statistics and electrodynamics was doomed from the outset to failure, and it was necessary to introduce a hitherto unknown discontinuity into the theory. It was, of course, clear that this "atomising of energy" would conflict sharply with existing and apparently well-founded theories. For if the energy of the *Planck* oscillator was only to amount to integral multiples of $\epsilon = h\nu$, and therefore was only to be able to have the values $0, \epsilon, 2\epsilon, 3\epsilon \ldots$ then, since the oscillator only changes its energy by emission and absorption, the conclusion was inevitable that oscillators cannot absorb and emit amounts of energy of any magnitude but only whole multiples of ϵ. (*Quantum emission and quantum absorption.*) This conclusion is in absolute contradiction to classical electrodynamics. For, according to the electron theory, an electromagnetic oscillator, for instance a vibrating electron, emits and absorbs in a field of radiation perfectly continuously, that is to say, in sufficiently short times it emits or absorbs indefinitely small amounts of energy.

§ 2. Einstein's Light-quanta ; Phenomena of Fluctuation in a Field of Radiation

Thus at the very entrance into the new country there yawned a gulf, which had either, in view of the previous success of the classical theory, to be bridged over by a compromise; or, failing this, tradition would have to be discarded and the gap would be relentlessly enlarged. *Einstein* felt him-

16

self compelled to take the latter radical course. On the basis
of very original considerations,[49] he set up the hypothesis that
the energy quanta not only played a part, as *Planck* held, in
the interaction between radiation and matter (resonators or
oscillators), but *that radiation, when propagated through a
vacuum or any medium, possesses a quantum-like structure
(Light-quantum hypothesis)*. Accordingly, all radiation was
to consist of indivisible "radiation quanta"; when the energy
is being propagated from the exciting centre, it is not divided
evenly in the form of spherical waves over ever-increasing
volumes of space, but remains concentrated in a finite number
of energy complexes, which move like material structures,
and can only be emitted and absorbed as whole individuals.
Einstein believed himself forced to this strange conception,
which breaks with all the observations that appear to
support the undulatory theory, by several investigations,
all of which led to the same conclusion. He was per-
suaded to this view by the result of calculations dealing
with certain phenomena of fluctuation, phenomena which
are familiar to us in statistics and particularly in the kinetic
theory of gases. It is well known that in a gas which
contains n molecules in a volume v_0, the spatial distribution
of these molecules is far from constant, being subject to variation on account of the motion of the molecules. Indeed, in
principle, extreme cases are possible as that, for example, in
which all n molecules are collected at a given moment in a
fractional part $v(< v_0)$ of the volume. The probability of
this rare constellation is known to be

$$w = \left(\frac{v}{v_0}\right)^n \qquad . \qquad . \qquad . \qquad . \qquad (26)$$

an extraordinarily small number when n is great; that is to
say, the event in question occurs extremely rarely.

Now, the spatial density of the radiation enclosed within a
volume v_0 is subject to quite analogous variations. If E is
the total energy of the radiation (supposed to be monochromatic) and if its frequency ν is so great, or its temperature
so low, that *Wien's* Law of Radiation holds for it, then the
probability that the whole radiation occupies the partial
volume $v(< v_0)$ is, according to *Einstein*,[50]

2

$$w = \left(\frac{v}{v_0}\right)^{\frac{E}{h\nu}} \quad . \quad . \quad . \quad . \quad (27)$$

A comparison with (26) shows that *the radiation, within the limits of validity of* Wien's *Law, behaves as if it were made up of* $n \left(= \dfrac{E}{h\nu} \right)$ *independent complexes of energy, each of magnitude* $h\nu$.

Two other investigations [51] of *Einstein* led to the same conclusion. In the first, a very large volume filled with black-body radiation is considered, which communicates with a small volume v. If E is the momentary energy of the radiation of frequency ν in the volume v, this energy varies, as is known, irregularly with the time about a mean value $\bar{E}$; the magnitude $\epsilon = E - \bar{E}$ is called the fluctuation of the energy. Now, the general theory of statistics leads to the following value [52] for the mean square, that is, for $\overline{\epsilon^2}$,

$$\overline{\epsilon^2} = kT^2 \cdot \frac{d\bar{E}}{dT} \cdot \quad . \quad . \quad . \quad (28)$$

If we replace $\bar{E}$ by the value obtained from *Planck's* Law of Radiation, we obtain for the mean square of fluctuation an expression with two terms,[53] in which only one term can be calculated on the basis of the classical undulatory theory; the second, which greatly exceeds the first in magnitude when the density of radiant energy is low (that is, at high frequencies or at low temperatures, in short, when *Wien's* Law is valid), can only be understood when we again picture the radiation as composed of indivisible energy-quanta.

The second of *Einstein's* two investigations, to which we referred above, deals with the fluctuations of impulse which a freely movable reflecting plate is subjected to in a field of black-body radiation on account of the irregular fluctuations of the pressure of radiation. If, in addition, the plate is subjected to the irregular blows of gas-molecules, under the influence of which it executes Brownian movements, there must be equilibrium between the impulses which the molecules on the one hand, and the radiation on the other, impart to the plate. If, now, we assume *Planck's* Law to hold for the radiation, there again follows for the mean square of the variations in impulse due to the radiation an expression in two terms, only one of which is explained by the undulatory theory of light. The other term points to a quantum-like structure of the radiation, and this suggests the introduction of the light-quantum hypothesis.

§ 3. Transformation of Light-quanta into other Light-quanta or Electronic Energy

However strange this hypothesis appeared, it was not to be denied that it was capable of explaining simply and naturally a number of phenomena which completely baffled the undulatory theory. A very striking example of this is afforded by the laws of phosphorescence, investigated by *P. Lenard* and his co-workers, and especially by *Stokes'* Law. For if ν_p is the frequency of the phosphorescent light emitted, and ν_e the frequency of the light exciting phosphorescence, then, according to *Einstein's* conception,[54] one quantum $h\nu_e$ of the exciting radiation is changed through absorption by the atom of the phosphorescent substance into one quantum $h\nu_p$ of the light of phosphorescence. According to the principle of energy, we must have $h\nu_e \geq h\nu_p$, i.e. $\nu_e \geq \nu_p$. And this is *Stokes'* Law.

Further, another fact in the realm of phosphorescence phenomena speaks against the undulation hypothesis and in favour of that of light-quanta. According to the classical undulatory theory, all molecules of a phosphorescent body on which a light-wave impinges, should absorb energy from the wave, and thus all simultaneously become able to emit phosphorescent light. In reality, relatively only very few molecules are excited to phosphorescence at the same time, and only gradually, in the course of time, does the number of molecules excited increase. It would thus appear as if the light-wave falling on the phosphorescent body has not equal intensity along its whole front—as the classical theory assumes—but rather as if it consists of single energy-complexes thrown out by the source of light, so that the wave-point possesses, as it were, a "beady" structure, in which active portions (light-quanta) alternate with inactive gaps.

This conception of the "beady" wave-front had played a part before the advent of *Einstein's* hypothesis of light-quanta. *J. J. Thomson* [55] had tried to make use of it to explain the fact that, when a gas is ionised by ultra-violet light or Röntgen rays, only a relatively extremely small number of gas-molecules are ionised.   This is a phenomenon which is quite analogous to the above-named phenomenon of phosphor-escence; for these, too, according to *Lenard's* view, the excitation consists in the disjunction, through the agency of the radiation, of electrons from the molecules of the phosphorescent body, and these electrons attach themselves to " storage atoms."   On the return of these electrons to the parent molecules, energy is set free and sent out as phosphorescent light.   The ionisation of gases by ultra-violet light or Röntgen rays [56] is also capable of being explained naturally by the light-quantum hypothesis.   If we suppose with *Einstein*, that one light-quantum $h\nu$ is used up in ionising one molecule, then $h\nu > J$, where J is the work required to ionise one molecule, that is to say, to remove an electron from it. We have under consideration here a phenomenon which belongs to the great branch of *photo-electric phenomena*,[57] i.e. the liberation of electrons from gases, metals, and other substances by the action of light.   According to the hypothesis of light-quanta, in all these processes light-quanta are changed into kinetic energy of the electrons hurled off from the body. If we again adopt *Einstein's* standpoint, according to which one light-quantum $h\nu$ is transformed into the kinetic energy of one projected electron, we must have the following relation [58] for the energy of emission of the emitted electrons, each having a mass m :

$$\tfrac{1}{2}mv^2 = h\nu - P .\qquad .\qquad .\qquad . \quad (29)$$

This is called *Einstein's* Law of the Photo-electric Effect.   In this, P is the work that has to be done to tear the electron away from the atom, and to project it from the point at which it is torn from the atom up to the point at which it leaves the surface of the body.   For the energy of the emitted electrons we thus obtain a linear increase with the periodicity of the light which releases them.   This law, which many investigators have attempted to prove, with varying success, has recently been verified by *R. A. Millikan* [59] for the normal photo-electric effect [60] of the metals Na and Li with such a degree of accuracy that we can actually use this method for the exact determination of h. The value found by *Millikan*, $h = 6\cdot57 \times 10^{-27}$, is in good agreement with the value $h = 6\cdot548 \times 10^{-27}$ found by *Planck* from radiation measurements.

In an entirely similar manner as was used for the phenomena of phosphorescence, the phenomena of *fluorescence* in the regions of the Röntgen and visible radiations may be explained by the hypothesis of light-quanta. The investigations of *Ch. Barkla, Sadler, M. de Broglie*, and *E. Wagner* [61] have shown the following: if a body is inundated with Röntgen rays, and if the absorption of these rays by the body is measured whilst the hardness (i.e. the frequency ν_e) of the rays is varied, the absorption, as we pass from lower to higher ν_e, suddenly increases to a high value for a certain value of ν_e. At the same moment the body begins, at the expense of the energy absorbed, to emit a *secondary Röntgen radiation characteristic of the body itself* in the form of a line spectrum. It further appears that all lines emitted have a lower ν than that of the exciting radiation. As a matter of fact, the hypothesis of light-quanta requires that the radiation-quantum $h\nu$ of all rays emitted as secondary radiation should be smaller than the quantum $h\nu$ of the primary exciting rays. For example, the region of frequencies which serves to excite the "K-series" stretches from a sharply defined limit ν_k (the so called "edge of the absorption band") upwards towards higher frequencies; whereby ν_k is somewhat larger than the hardest known line (γ) of the K-series. In other words, the excitation of secondary Röntgen radiation by primary Röntgen rays also obeys *Stokes'* Law.

§4. The Transformation of Electronic Energy into Light-quanta

It is very significant, that the transformation of light-quanta into kinetic energy of electrons is also, as it were, "reversible," that is, the opposite process also occurs in nature, by which light-quanta result from the kinetic energy of charged particles. A good example of processes of this kind is afforded by the generation of Röntgen rays by the impact of quickly-moving electrons (cathode rays) on matter. If, say, the characteristic K-series of a certain element is to be generated by the impact of cathode rays upon an anti-cathode formed of the said element, then the kinetic energy E of an impinging electron must exceed a critical value E_K. For if we imagine E changed into a light-quantum $h\nu_e$, then ν_e must fall within the region of excitation of the K-series, and must thus be $\geq \nu_K$ (ν_K being the frequency of the edge of the absorption band). It follows that $E \geq h\nu_K (= E_K)$. From this there follows an important relation between the frequency ν_K of the edge of the absorption band and the critical value E_K of the electronic energy, i.e. the smallest value of the energy at which the electron is just able to generate the required secondary radiation. This quantum-relation $E_K = h\nu_K$ has proved quite correct according to measurements carried out by *D. L. Webster* [62] and *E. Wagner*,[63] and conversely presents, when E_K and ν_K are sufficiently accurately known, a method for the determination of h.[64]

Now, it is known that the cathode rays, on striking the anti-cathode, do not merely excite the characteristic Röntgen radiation, that is a line spectrum, but excite a continuous spectrum as well, the so-called " impulse radiation " (*Bremsstrahlung*). If we therefore select any frequency ν of this continuous spectrum, the ideas of the hypothesis of light-quanta immediately suggest the conclusion that a definite minimum energy E_m of the impinging electrons is necessary to excite this frequency ν, and that we must have $E_m = h\nu$. The investigations of *D. L. Webster*,[62] *W. Duane* and *F. L. Hunt* [65], *A. W. Hull* and *M. Rice*,[66] *E. Wagner*,[67] *F. Dessauer* and *E. Back* [68] have confirmed these formulæ with the greatest accuracy, and thus form the foundation of one of the most trustworthy methods for the precise measurement of the magnitude h. The following values were obtained: $h = 6\cdot50 \times 10^{-27}$ (*Duane-Hunt*); $h = 6\cdot53 \times 10^{-27}$ (*Webster*); $h = 6\cdot49 \times 10^{-27}$ (*Wagner*).

We also meet with similar phenomena in the visible and neighbouring regions of the spectrum. Thus *J. Franck* and *G. Hertz* [69] showed that the impact of electrons upon mercury vapour molecules can be used to excite a definite characteristic fluorescence line of mercury of wave-length $\lambda_0 = 2536 \overset{\circ}{A}$ (i.e. $\nu_0 = 1\cdot183 . 10^{15}$), if the kinetic energy of the electron exceeds a certain critical value E_0. In this connexion they found that the relation $E_0 = h\nu_0$ was again fulfilled with great accuracy.[70]   We shall return to these experiments and others connected with them later, since they play an important part in confirming the most recent model of the atom.

§ 5. Other Applications of the Hypothesis of Light-quanta

In a considerable number of other cases, which shall only be noticed shortly at this point, the hypothesis of light-quanta has proved of value, especially in the hands of *J. Stark*[71] and *Einstein*.   Thus *Stark*[72] has made use of this hypothesis to interpret the fact that the canal-ray particles emit their "kinetic radiation" only when their speed exceeds a certain value.   He has also propounded general laws for the position of band-spectra of chemical compounds by arguing on the basis of the hypothesis of light-quanta.[73]   Finally, *Einstein*[74] and *Stark*[75] have considered photo-chemical reactions from the standpoint of the hypothesis of light-quanta and have enunciated a fundamental law, which has been verified, at least partially, by the detailed investigations of *E. Warburg*.[76]

§ 6. Planck's Second Theory

In spite of all the successes which the quantum hypothesis of light is able to show, we must not leave out of consideration that this radical view, at least in its existing form, is very difficult to bring into agreement with the classical undulatory theory.   Since on the one hand the phenomena of interference and diffraction, in all their observed minutiæ, are excellently described by the wave-theory, but offer almost insuperable difficulties to the quantum theory of light, it is easy to understand that few scientists could make up their minds to approve of such a far-reaching change in the old and well-tested conception of the propagation of light, a change that entailed perhaps its complete abandonment.   This more cautious and conservative standpoint was taken up by *M. Planck*, who retains it to this day, inasmuch as he preferred to locate the quantum property in matter (the oscillators)—or at least to confine it to the process of interaction between matter and radiation—while endeavouring to retain the classical wave-theory for the propagation of radiation in space. None the less, serious hindrances had already intruded themselves in the development of his first quantum hypothesis (quantum emission and quantum absorption). For *H. A. Lorentz*[77] pointed out quite rightly that the conception, especially of quantum absorption, leads to peculiar difficulties. He showed that the time which an oscillator requires for the absorption of a quantum of energy turns out to belong to an improbable degree when the external field of radiation is sufficiently weak. Moreover, it would be possible to interrupt the radiation at will before the oscillator had absorbed a whole quantum. As a result of these objections *Planck* determined to modify the quantum hypothesis as follows.[78] *Absorption proceeds continuously and according to the laws of classical electrodynamics : the energy of the oscillators is therefore continuously variable, and can assume any value between* 0 *and* ∞ . *On the other hand, emission occurs in quanta, and the oscillator can emit only when its energy amounts to just a whole multiple of* $\epsilon = h\nu$. *Whether it then emits or not is determined by a law of probability. But if it does emit, then it loses its whole momentary energy, and therefore emits quanta. Between two emissions its energy-content grows by absorption continuously and in proportion to the time.*

According to this second theory of *Planck*, which is called the theory of quantum emission, the mean energy U of a linear oscillator is $\frac{h\nu}{2}$ greater than in the first theory.[79] While in the former case the mean energy of the oscillator at absolute zero was equal to zero (see equation (9) from which, when $T = 0$, $\bar{U} = 0$), in the case of this second theory it is equal to $\frac{h\nu}{2}$. The oscillators retain therefore at the zero-point a zero-point energy of value $\frac{h\nu}{2}$ as a mean, inasmuch as they assume, when $T = 0$, all possible energies between 0 and $h\nu$. Nevertheless, this theory also, when the relation (7) is correspondingly modified, leads to *Planck's* Law of Radiation.

In the course of time *Planck* has made several further

attempts [80] to enlarge and modify this second theory too. For example, he has temporarily assumed the emission also to be continuous, and relegated the quantum element to the excitation of the oscillators by molecular or electronic impacts. He has, however, repeatedly returned in essentials to the second form of his theory (continuous absorption, quantum emission).

§ 7. Zero-point Energy

In more than one direction, this theory has had further results. The appearance of the mean zero-point energy, which is peculiar to this second theory of *Planck*, became the starting-point of a series of researches, in which certain physicists, going beyond *Planck*, postulated the existence of a *true* (not *mean*) zero-point energy equal for all oscillators. On this basis, *Einstein* and *O. Stern* [81] have given a deduction of *Planck's* Law which avoids all discontinuities other than the existence of this zero-point energy.

In the year 1916, *Nernst* [82] took a still more radical step in postulating the existence of a "zero-point radiation" which was also to be present at the absolute zero of temperature and was to exist independently of heat radiation, filling the whole of space, and such that the oscillators, as well as all molecular structures, set themselves in equilibrium with it by taking up the zero-point energy. Even if we regard these views more or less sceptically, one thing cannot be ignored : many facts undoubtedly support the conception that at the absolute zero by no means all motion has ceased. We need only draw attention to the fact, that, according to the view of *F. Richarz*,[83] *P. Langevin*,[84] and according to the experiments of *Einstein*, *W. J. de Haas* [85] and *E. Beck*,[86] Para- and Diamagnetism are produced by rotating electrons and that this magnetism remains in existence down to the lowest temperatures.

§ 8. Theory of the Quantum of Action

In yet another respect has *Planck's* theory proved stimulating, in virtue of a special formulation which *Planck* gave it [87] at the Solvay Congress in Brussels during 1911. For here *Planck* gave expression for the first time to the idea that the appearance of energy-quanta is only a secondary matter, being only the consequence of a deeper and more general law. This law, which is to be regarded as the precursor of the latest development of the doctrine of quanta, may be formulated as follows: Suppose the momentary state of a *Planck* oscillator, say a linearly vibrating electron, to be defined according to *Gibb's* method by its *displacement q* from its position of rest and by its *impulse* or *momentum p*, and suppose it to be represented in a *q-p* plane (the state- or phase-plane). Every point of the *q-p* plane, that is, every *phase-point*, corresponds to a definite momentary condition of the oscillator. The postulate is then made that not all points of this plane of states are equivalent. *On the contrary, there*

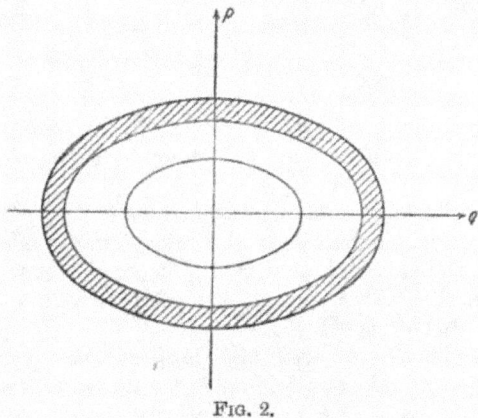

Fig. 2.

are certain states of the oscillator which are distinguished by a peculiarity. The totality of the phase-points that correspond to these peculiar states form a family of discrete curves which surround one another. In the case of the *Planck* oscillator these curves are concentric ellipses (see Fig. 2) which divide the phase-plane into ring-like strips. The postulate of the quantum theory now consists in this, that these ring strips all possess the same area *h*. If we calculate on this basis the energy possessed by an oscillator in one of these unique states, we find [80] that it is a whole multiple of *hv*. These special states (represented in the phase-plane by the points of the discrete ellipses) are, therefore, according to *Planck's* first theory, the only dynamically possible and stable states of the oscillator. If an oscillator emits or absorbs, its phase-point jumps from one ellipse to another. The state of affairs is different if we accept *Planck's* second theory. According to this, *all* conditions of the oscillator, that is *all* points on the phase-plane, are dynamically possible. On the other hand, emission takes place only in the states specially distinguished by the ellipses. Seen from this new point of view, the energy-quanta are, therefore, only a result of the partitioning of the phase-plane. Mathematically, we may express this " structure of the phase-plane " thus : the *n*th unique curve encloses a surface of area *nh*, or, in symbolic language,

$$\iint dqdp = \int pdq = nh \qquad . \qquad . \qquad . \quad (30)$$

The double integral is taken over the surface; the single integral is taken around the boundary curve of the *n*th ellipse.

On this basis for systems of one degree of freedom, which is called *Planck's* theory of " the action-quantum "—for *h* has the dimensions of an action—the modern extension of the quantum theory for several degrees of freedom has, as we shall see, been erected.

Further, a line of argument proposed and developed by *A. Sommerfeld* takes its origin here. Starting from the fact just mentioned, that *Planck's* constant *h* possesses the dimensions of action (energy-time), *Sommerfeld* set up the hypothesis [89] that for every purely molecular process, say the release of an electron in the photo-electric effect, or the stopping of an electron by the anticathode in the generation of Röntgen rays, the *quantity* called *action* $\int_0^\tau (L - V)dt$, known to us from *Hamilton's* Principle, has the value $\frac{h}{2\pi}$. Here *L* and *V* are the kinetic and potential energies of the electron respectively, τ is the duration of the molecular process, say, for example, the time which is required for the release of the electron from the atomic complex during the photo-electric effect, or the stopping of the electron by the anti-cathode.

This formulation of the quantum hypothesis is, as it were, an expression of the well-known fact that large amounts of energy are absorbed or given up in short times, whereas small amounts are absorbed or emitted in longer times by the molecules, so that on the whole the product of the energy transferred and the duration of the time of exchange is a constant. In fact, fast cathode rays, for example, are stopped by matter in a shorter time—and therefore generate harder Röntgen rays—than slow cathode rays. *Sommerfeld* has applied his theory successfully to the mechanism of the generation of Röntgen rays and γ-rays.[90] *Sommerfeld* and *P. Debye*[91] have worked out on the same basis a theory of the photo-electric effect, which, like the hypothesis of light-quanta, also leads to *Einstein's* Law (29).

CHAPTER IV

The Extension of the Doctrine of Quanta to the Molecular Theory of Solid Bodies [92]

§ 1. Dulong and Petit's Law

IT was a particularly fortunate circumstance for the consolidation of the doctrine of quanta that the failure of classical statistics was not confined to the theory of radiation, but, as appears later, extended to the molecular theory of solid bodies. Thus there arose in quite another field a strong support for the quantum hypothesis, namely, in the field of *Atomic Heats*. The Atomic Heat of a substance (in the case of polyatomic bodies we say the " Molecular Heat ") is defined as the product of its specific heat and its atomic weight (or molecular weight) ; or, otherwise expressed, it is that amount of heat which must be communicated to a " gramme-atom " [93] (or gramme-molecule) of the body, in order that its temperature may be raised by one degree. According to our present conceptions, the thermal content of a monatomic solid, say a crystal, is nothing more than the energy of the elastic vibrations of its atoms, which are arranged in the form of a space-lattice, about their positions of equilibrium. If we apply classical statistics to these vibrations, and particularly the law of equipartition of kinetic energy, we arrive at the following conclusion : The mean kinetic energy of an atom vibrating in space, i.e. with three degrees of freedom, is $\frac{3kT}{2}$, and its mean potential energy is equal to the same amount,[94] so that its total energy is therefore $3kT$. If we now consider 1 gramme-atom of the body, that is, a system of N atoms (where N is the *Avogadro* number, approximately 6×10^{23}), we get for the mean energy of the body, remembering (19),

$$\bar{E} = 3kTN = 3RT \qquad . \qquad . \qquad . \quad (31)$$

where R is the absolute gas-constant. It follows that the atomic heat of the body at constant volume becomes:

$$C_v = \frac{d\bar{E}}{dT} = 3R = 5\cdot94 \left[\frac{\text{cal.}}{\text{deg.}}\right] \quad . \quad . \quad (32)$$

This is the law of *Dulong* and *Petit*,[95] according to which *the atomic heat (at constant volume) of monatomic solid bodies has the value* $5\cdot94 \frac{\text{cal.}}{\text{deg.}}$, *independently of the temperature.*[96] This law is actually obeyed by many elements more or less closely.[97] On the other hand, elements have long been known which are far from following this rule, and which show systematic differences, especially at low temperatures.

Thus, as early as the year 1875, *F. H. Weber*[98] found that the atomic heat of diamond at $-50°$ C. is about $0\cdot75 \frac{\text{cal.}}{\text{deg.}}$. The atomic heats of other elements as well (boron, beryllium, silicon) have also been shown to be much too small at ordinary temperatures. And altogether it appeared that the defect from *Dulong* and *Petit's* normal value occurs quite generally at low temperatures, and becomes the more pronounced, the lower the temperature. The classical theory offered no solution of these low values of the atomic heat.[99]

§ 2. Einstein's Theory of Atomic Heats

Einstein was the first to recognise [100] that in this case, too, the quantum theory was destined to solve the difficulty. Precisely as in the theory of radiation, the method of classical statistics leads of necessity to a wrong law in the field of atomic heats. *Hence, here also, we must abandon the law of the equipartition of energy.* In fact, we need only imagine electric charges distributed among the atoms [101] and then we see that, exactly like the *Planck* oscillators, they must set themselves in equilibrium with the heat-radiation which is always present in the body. This means, however, that the relation (7), according to which $\bar{U} = \frac{c^2}{v^3} \mathbf{K}_v$, must be set up between the mean energy $\bar{U}$ of an atom vibrating linearly with frequency v, and the intensity of radiation $\mathbf{K}_v$. If we now take *Planck's* radiation formula (12) as empirically given, it follows immediately that the mean energy $\overline{U}$ of the *linearly* vibrating atom must possess, not the value kT given by classical statistics, but the value given by the quantum theory, namely, $\overline{U} = \dfrac{h\nu}{e^{\frac{h\nu}{kT}} - 1}$. For the atom which vibrates in space we get, therefore—by an obvious generalisation—in place of the classical value $3kT$, the quantum value: $\dfrac{3h\nu}{e^{\frac{h\nu}{kT}} - 1}$.

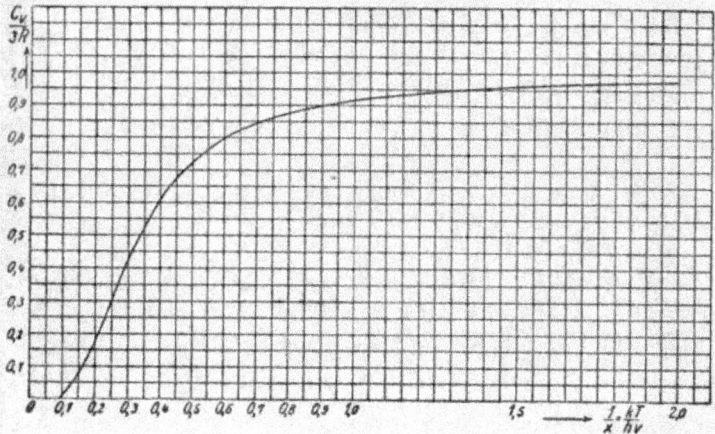

FIG. 3.

The heat-content of the gramme-atom will therefore be

$$\overline{E} = \frac{3Nh\nu}{e^{\frac{h\nu}{kT}} - 1} \qquad . \qquad . \qquad . \qquad . \quad (33)$$

from which we get for the atomic heat at constant volume *Einstein's* formula

$$C_v = \frac{d\overline{E}}{dT} = 3R \cdot \frac{x^2 e_x}{(e^x - 1)^2}, \quad \text{where} \quad x = \frac{h\nu}{kT} \quad . \quad (34)$$

According to this, the atomic heat of monatomic solid bodies is not a constant which is independent of the temperature, as Dulong *and* Petit's *Law requires, but is a function of* $\frac{\nu}{T}$, *and*

*is therefore in the case of a definite body (i.e. with ν fixed)
a function of the temperature.* Its form is such (see Fig. 3),
that for $T = 0$ (i.e. $x = \infty$) the atomic heat itself is
zero, and then increases gradually with increasing temperature, approaching asymptotically at high temperatures
(i.e. with small x) the classical value $3R$. *Dulong* and
Petit's Law is therefore only true in the limit for small
values of $\dfrac{h\nu}{kT}$, that is, for low frequencies of atomic vibration,
or high temperatures, exactly as is the case with *Rayleigh's*
Law of Radiation. The departures from *Dulong* and *Petit's*
Law, in passing from high to low temperatures, become marked
the sooner the greater the frequency of the atoms.

§ 3. Methods of Determining the Frequency

This frequency ν—the only unknown magnitude in *Einstein's*
formula (34)—may be determined by several independent and
very noteworthy methods. One way that is always possible
is of course the following: For a given substance we choose
an experimentally well-known value of the atomic heat C_v^*,
which corresponds to a definite temperature T^*. From (34)
it follows then that $\dfrac{x^2 e^x}{(e^x - 1)^2} = \dfrac{C_v^*}{3R}$, an equation from which
$x = \dfrac{h\nu}{kT^*}$ can be determined, and thence ν. From the ν thus
found the course of the whole C_v curve can be calculated for
all temperatures, and compared with experiment.

Besides this "empirical" method of determining ν, there
are a number of other more "theoretical" methods which
do not require the use of the values of the atomic heat.
Einstein,[102] as far back as 1911, discovered an important
connection between the frequency ν and the elastic properties
of the body. That such a connexion must exist is easily
recognised from the following considerations: imagine the
atoms of the body arranged upon a space-lattice, as in a
crystal, and suppose a certain definite atom arbitrarily disturbed from its position of rest, then this atom, when released,
will execute vibrations about its position of equilibrium. If
we suppose these vibrations to be simply periodic ("monochromatic")—we shall, however, soon recognise that this

supposition is an inadmissible approximation—we see that the frequency ν is the greater the smaller the atomic mass, and therefore also the atomic weight of the body, and the greater on the other hand the force which restores the atom to its position of equilibrium. This restoring force is, however, for its part the stronger, the less extensible and therefore compressible the body is. Hence ν must turn out the greater, the smaller the atomic weight and the compressibility of the substance. The exact working out of this idea led *Einstein* to the formula [103]

$$\nu = \frac{2 \cdot 8 \cdot 10^7}{A^{\frac{1}{3}} \rho^{\frac{1}{6}} \kappa^{\frac{1}{2}}} \quad . \quad . \quad . \quad . \quad (35)$$

Where A is the atomic weight, ρ the density, and κ the compressibility of the body.

A further interesting relation, which connects ν with thermal data, namely, the melting-point, was found by *F. A. Lindemann* [104] by working out the conception that the amplitude of vibration of the atom at the melting-point is of the order of magnitude of the distances between the atoms. If T_S is the absolute melting-point, then it follows that

$$\nu = 2 \cdot 8 \cdot 10^{12} \cdot \frac{T_S^{\frac{1}{2}} \rho^{\frac{1}{6}}}{A^{\frac{5}{6}}} \quad . \quad . \quad . \quad (36)$$

Another formula deduced by *E. Grüneisen* [105] may also be given here:

$$\nu = 2 \cdot 91 \cdot 10^{11} \cdot A^{-\frac{5}{6}} \left[C_v^{\frac{1}{2}} \cdot a^{-\frac{1}{2}} \cdot \rho^{\frac{1}{6}} \right]_0 \quad . \quad . \quad (37)$$

Here C_v is the atomic heat at constant volume, and a is the coefficient of thermal expansion; the index 0 means that the value of $C_v^{\frac{1}{2}} a^{-\frac{1}{2}} \rho^{\frac{1}{6}}$ at absolute zero is to be used.

From formulæ (35) and (36) we recognise at once the abnormal behaviour of diamond, for example, in respect to its atomic heat. For it is known that diamond has a high melting-point and very low compressibility accompanied by a low atomic weight. Its ν is therefore comparatively large, and it follows therefore, according to the above considerations, that its atomic heat falls below *Dulong* and *Petit's* value of $3R = 5 \cdot 94 \frac{\text{cal.}}{\text{deg.}}$ at comparatively high temperatures. In fact

3

the atomic heat of diamond at 284° abs. is only $1.35 \frac{\text{cal.}}{\text{deg.}}$, at

413° abs. it is $3.64 \frac{\text{cal.}}{\text{deg.}}$, and even at 1169° abs. it reaches only

the value $5.24 \frac{\text{cal.}}{\text{deg.}}$.

Finally, particular importance attaches to a relation, first discovered by E. Madelung [106] and W. Sutherland,[107] between the frequency ν of the atoms and the optical properties of bodies. The two investigators started in this case from the following conception : Crystals of diatomic compounds (binary salts), such as rock-salt (NaCl), sylvin (KCl), potassium bromide (KBr), and others, are known to be cubical space-lattices, in which the single atoms carry electric charges, and therefore appear as ions. In fact, the points of the space-lattice are occupied alternately by the positively charged Na^+ (or K^+) atoms, and the negatively charged Cl^- (or Br^-) atoms. If an electromagnetic light-wave of frequency ν falls upon this crystal, the two ions are thrown into forced oscillations relatively to one another, and further, on account of "resonance," the more strongly, the more exactly the frequency ν of the impinging wave agrees with the natural frequency ν_r, which lies in the infra-red, of the ions themselves. Since the ionic vibrations are set up at the cost of the energy of the impinging wave, this energy will be weakened (absorbed) the more during its passage through the body, the nearer ν lies to ν_r. On the other hand, the vibrating ions radiate back waves of frequency ν since they are compelled to execute these vibrations, when set into forced vibration, doing so the more strongly, the more pronounced the resonance is, again, therefore, the nearer ν lies to ν_r. Hence a region of maximum absorption and strongest (metallic) reflection will lie in the neighbourhood [108] of $\nu = \nu_r$. These regions of metallic reflection of a given substance may be detected by the method of "Reststrahlen" (residual rays) worked out by H. Rubens and E. F. Nichols.[109] For this purpose we only require to reflect radiation of a considerable range of frequency about ν repeatedly from the substance. In this way all waves will be gradually absorbed except those most strongly reflected. These are, however, just those of frequency v_r. They are thus "residual." The ultra-red frequency v_r of the ions therefore agrees with the frequency of the residual rays.[110] On the other hand, this vibration of the charged atoms is dependent on the elastic properties of the substance, as we recognised in considering the formula (35). We thus conclude that the "elastic" frequency of the atoms of binary salts agrees to a close approximation with the "optical" frequency of their residual rays. But since the "elastic" frequency of the atoms determines the behaviour of their atomic heat, the ring is thereby closed, and W. Nernst [111] was thus justified in propounding the fundamental law, *that in calculating the atomic heat of binary salts, we may simply insert for the atomic frequencies v the frequencies of the residual rays.*

In this way a number of independent ways were opened up for determining the atomic frequencies required for the calculation of the atomic heat. A comparison of the various values of v determined by these different methods shows in general satisfactory agreement, at any rate in order of magnitude.[112] One could hardly expect more, as we shall soon see, in view of the many idealised conditions that were used in the theory.

§4. Nernst's Heat Theorem

With a view to discovering experimentally the general law for the decrease of the atomic heat when approaching low temperatures W. Nernst [113] began in 1910, in co-operation with his research students, a series of masterly and widely planned researches. For, by an entirely different route from *Einstein* —namely, by way of thermodynamics—he also had become convinced that the atomic heat of solid bodies must become vanishingly small on approaching absolute zero. In his opinion this result was only one of several consequences of a general principle, namely, a new law of heat.[114] This Heat Theorem of *Nernst*—often called the Third Law of Thermodynamics—states, in its original form, the following fact: If we regard a system of condensed (i.e. liquid or solid) bodies, which passes at temperature T by means of an isothermal reaction from one state to another, and if A is the maximum work which can be gained from this reaction, then

$$\frac{dA}{dT} = 0 \text{ for the limit } T = 0 \qquad . \qquad . \quad (38)$$

that is to say, in the immediate neighbourhood of absolute zero, the maximum work which can be gained is independent of the temperatur. But it follows immediately from this, if we apply the two laws of thermodynamics,[115] that for any reaction which changes the system from the initial condition with energy U_1 to the final condition with energy U_2, the relation holds that

$$\frac{dU_1}{dT} = \frac{dU_2}{dT} \text{ for the limit } T = 0 \qquad . \qquad . \quad (39)$$

Now, since $\frac{dU}{dT}$, if we take a gramme-atom of the substance, gives the atomic heat, we are led to enunciate the following rule: *in the immediate neighbourhood of absolute zero, the atomic heat of condensed systems remains unchanged during any transformation.*

Planck [116] has given *Nernst's* Theorem a still more general form: *Not only the difference of the atomic heats (before and after the reaction) is to assume the value 0 at absolute zero, but also each atomic heat itself is to do the same.* Thus it follows from the extended *Nernst* Theorem, in agreement with the demands of the quantum theory, that the atomic heats of solid bodies disappear at absolute zero.

§ 5. The Improvement on Einstein's Theory of Atomic Heats

The experiments of *Nernst* and his collaborators proved quite convincingly that the atomic heat of all solid bodies tends towards a zero value as the temperature falls. In the main, the courses of these decreasing values showed a notable agreement with *Einstein's* formula (34). At low temperatures, however, systematic discrepancies were found in all cases, in the sense that the observed atomic heats fell off much more slowly than *Einstein's* formula demanded.[117] *W. Nernst* and *F. A. Lindemann* [118] tried to take these discrepancies into account by constructing an empirical formula, and this actually expressed the observations much more accurately than did the *Einstein* formula. This *Nernst*-

Lindemann formula, which is now only of historical interest, is as follows :—

$$C_v = \frac{3R}{2}\left\{ \frac{x^2 e^x}{(e^x - 1)^2} + \frac{\left(\frac{x}{2}\right)^2 e^{\frac{x}{2}}}{\left(e^{\frac{x}{2}} - 1\right)^2} \right\}, \quad \text{where } x = \frac{h\nu}{kT} \quad . \quad (40)$$

It receives a meaning if we suppose that one half of all the atoms vibrate with the frequency ν, the other half with the frequency $\frac{\nu}{2}$. While this supposition is untenable in this raw form, it contains a kernel of truth, namely, recognition of the fact that the "monochromatic" theory of atomic heats, which assumes only a single fixed frequency ν for all atoms, goes too far, being an idealisation of the real state of affairs. *Einstein*, who at first, for the sake of simplicity, reckoned with only one frequency, had himself already recognised how matters stood, and drawn attention to the need for amending his theory.[119] Nowadays, in fact, we think of a solid body, say a crystal, as built up of atoms regularly arranged upon a space-lattice, according to *Bravais'* conception ; and this hypothesis has been verified as a certainty through *Laue's* discovery of the interference of Röntgen rays. In such a complicated mechanical system, however, the single atoms do not vibrate independently of one another with a single frequency ν. But the position of equilibrium of each atom, and thereby the type of its oscillations about that position, is determined rather by the forces which all the other atoms of the body exert upon the atom in question. We are confronted with a structure which is comparable to the one-dimensional case of a vibrating string, and which thus possesses a whole spectrum of natural frequencies, corresponding to the overtones of the string. If the body consists of N atoms, it possesses in general $3N$ natural frequencies,[120] of which the slowest are sound waves, while the quickest fall in the infra-red. The most general possible movement of each atom then consists in a super-position of all these natural frequencies. Now, since each natural frequency represents a linear, i.e. simple periodic, motion, exactly like the motion of a *Planck* oscillator, the idea naturally suggested itself, in calculating the energy-content of the body, to allot to each natural frequency of period ν the theoretical quantum amount $\dfrac{h\nu}{e^{\frac{h\nu}{kT}} - 1}$ as if the natural period were identical with a linear oscillator. The total mean energy of the body then becomes

$$\bar{E} = \sum_{i=1}^{3N} \frac{h\nu_i}{e^{\frac{h\nu_i}{kT}} - 1} \qquad . \qquad . \qquad . \quad (41)$$

in which the summation is carried over all $3N$ natural frequencies $\nu_1, \nu_2, \nu_3, \ldots \nu_{3N}$, that is, over the whole elastic spectrum of the substance. By differentiation with respect to T we obtain the atomic heat

$$C_v = \frac{d\bar{E}}{dT} = k \sum_{i=t}^{3N} \frac{x_i^2 e^{x_i}}{(e^{x_i} - 1)^2} \text{ where } x_i = \frac{h\nu_i}{kT} \quad . \quad (42)$$

§ 6. Debye's Theory of Atomic Heats

The kernel of the problem thus consists in calculating the "elastic spectrum" of a given body, that is, in determining for any body the position of its natural periods. In this sense, the theory has been worked out from two different sides; on the one hand by P. Debye,[121] who took an elastic continuum as an approximation to the actual atomically constructed body, and on the other by M. Born and v. Kármán,[122] who replaced the crystal of limited size by one of infinite dimensions. The difference between these two methods of approximation causes the main problem, namely, the working-out of the elastic spectrum, to be solved quite differently in the two cases. The Debye theory, which from the outset leaves out of consideration the crystalline, and even the atomic, structure of the body, rests upon the classical theory of elasticity, which, of course, treats bodies as structureless continua. From it follows the important law: the number $Z(\nu)d\nu$ of all those natural periods, the frequency of which falls within the interval $\nu, \nu + d\nu$, amounts to [123]

$$Z(\nu)d\nu = 4\pi V\left(\frac{1}{c_l^3} + \frac{2}{c_t^3}\right)\nu^2 d\nu \qquad . \qquad . \quad (43)$$

Here V is the volume of the body, c_l and c_t are the velocities with which longitudinal and transverse waves, respectively, are propagated within the body. In this case, however, the following difficulty occurs in replacing the body, which in reality consists of N atoms, by a continuum, namely, the elastic spectrum extends to infinity, that is, the number of natural frequencies becomes infinitely great. For example, the number of natural frequencies (fundamental tone and over-tones) of a linear string of length L are

$$\nu_i = c_t \cdot \frac{i}{2L} \text{ and } \nu_i = c_l \cdot \frac{i}{2L} \text{ respectively } (i = 1, 2, \ldots \infty)$$

according as to whether we are considering transverse or longitudinal frequencies. The series of overtones therefore extends without limit to infinity. In reality, however, as the body consists of N atoms (mass-points), it may not possess more than $3N$ natural frequencies. In order to attain this, *Debye* helps himself out by means of the following bold supposition. Instead of calculating strictly the elastic spectrum of the real body consisting of N atoms, he replaces it by that of the continuum as an approximation, *but breaks it off arbitrarily at the 3Nth natural period.* *Debye* thus gets the greatest frequency ν_m which occurs, that is, the upper limit of the elastic spectrum, from the condition :

$$\left.\begin{array}{c} \displaystyle\int_0^{\nu_m} Z(\nu)d\nu = \frac{4\pi V}{3}\left(\frac{1}{c_l^3} + \frac{2}{c_t^3}\right)\nu_m^3 = 3N \\[4mm] \text{therefore} \\[2mm] \displaystyle \nu_m = \left[\frac{9N}{4\pi V\left(\dfrac{1}{c_l^3} + \dfrac{2}{c_t^3}\right)}\right]^{\frac{1}{3}} \end{array}\right\} \qquad . \quad (44)$$

The atomic heat of the body, which follows from (42), is

$$C_v = k \cdot \int_0^{\nu_m} \frac{\left(\dfrac{h\nu}{kT}\right)^2 \cdot e^{\frac{h\nu}{kT}}}{\left(e^{\frac{h\nu}{kT}} - 1\right)^2} \cdot Z(\nu)d\nu$$

a result which can easily be brought into the following more simple form : [124]

$$C_v = \frac{9R}{x_{m3}} \cdot \int_0^{x_m} \frac{x^4 e^x dx}{(e^x - 1)^2}, \quad \text{where} \quad x_m = \frac{h\nu_m}{kT} = \frac{\Theta}{T} \quad . \quad (45)$$

The atomic heat is therefore only a function of the magnitude x_m, that is, it depends only on the ratio $\frac{\Theta}{T}$: here $\frac{\Theta}{T} = \frac{h\nu_m}{kT}$.

This result may be expressed in *Debye's* terms thus: *reckoning the temperature T as a multiple of a temperature Θ which is characteristic of the particular body, then the atomic heat is represented for all monatomic bodies by the same curve.* Hence we must be able to bring the C_v curves of all monatomic bodies into coincidence, if only the scale of temperature be suitably chosen for each substance.[125] For high temperatures, the *Debye* formula passes over, as it must do, into the classical value of *Dulong* and *Petit,* $C_v = 3R,$[126] just as do the *Einstein* and *Nernst-Lindemann* formulæ. On the other hand, it differs from these latter in falling much more slowly at low temperatures. For while the atomic heats, according to both *Einstein* and *Nernst-Lindemann*, fall exponentially $\left(\text{with } \frac{1}{T^2} \cdot e^{-\frac{\text{const}}{T}}\right)$ at low temperatures, *Debye's* formula leads to the fundamental law,[127] *that the atomic heats of all bodies at low temperatures are proportional to the third power of the absolute temperature.*

It is further remarkable, that we may write formula (44) for the maximum natural frequency in a form such that only measurable magnitudes occur in it. For if we express the two velocities of sound c_t and c_l in terms of the elastic constants of the body, and replace the volume V of the gramme-atom by the quotient $\dfrac{\text{atomic weight (A)}}{\text{density } (\rho)}$, it follows that [128]

$$\nu_m = \frac{5 \cdot 28 \cdot 10^7 \cdot \psi(\sigma)}{A^{\frac{1}{3}} \rho^{\frac{1}{2}} \kappa^{\frac{1}{2}}}$$

$$\text{where} \quad \psi(\sigma) = \left\{ \frac{2}{3}\left[\frac{2(1 + \sigma)}{3(1 - 2\sigma)}\right]^{\frac{3}{2}} + \frac{1}{3}\left[\frac{1 + \sigma}{3(1 - \sigma)}\right]^{\frac{3}{2}} \right\}^{-\frac{1}{3}} \quad (46)$$

In it κ is again the compressibility of the body, σ the *Poisson* ratio, that is, the ratio of the transverse contraction to the extension. The similarity of this formula with the *Einstein* relation (35) strikes one immediately. But in this case the second elastic constant of the isotropic body, σ, enters into the equation as well. Altogether, the upper limit ν_m of the elastic spectrum, at which, as one can show,[129] the natural frequencies always crowd together closely, plays in the stricter theory an analogous rôle to that played by the single natural frequency ν in the "monochromatic" theory.

Comparison with experiment shows[130] that the *Debye* formula, at any rate for the monatomic elements such as aluminium, copper, silver, lead, mercury, zinc, diamond, describes the course of values of the measured atomic heats very accurately. Particularly at low temperatures, the proportionality between the atomic heat and the third power of the absolute temperature receives fair confirmation.[131] In view of the fact that the idealised view (replacement of the actually atomic body by a continuum) is carried very far, we must not regard the agreement between theory and experiment as self-evident. At low temperatures, *Debye's* idealisation will justify itself. For then $\dfrac{h\nu}{kT}$ is large, and hence the amount of energy $\dfrac{h\nu}{e^{\frac{h\nu}{kT}} - 1}$ is small, excepting when ν itself assumes small values. *At low temperatures, therefore, only long waves will contribute sensibly to the energy of a body, and hence to its atomic heat.* For long waves, however, that is, for waves, the length of which is great compared with the distance between the atoms, the specific atomistic construction of the body plays no part; for them the substance is almost a continuum. The position is quite different at high temperatures, at which the longer frequencies up to the maximum ν_m (that is, the shorter waves down to the smallest) furnish contributions of energy. For the waves which correspond to the highest frequencies possess lengths, as can easily be shown,[132] which are comparable with the distances between the atoms, and for these shorter waves the medium cannot fail to betray its atomic structure. Here, therefore, its replacement by a continuum becomes questionable since the approximation is only very rough.

§ 7. The Lattice Theory of Atomic Heats according to Born and Kármán. The Elastic Spectrum of the most general Crystal

At this point the above-mentioned investigations of *Born* and *Kármán* intervene, which, going beyond *Debye*, take account of the real crystalline structure of the body, that is to say, the space-lattice arrangement of the atoms. In order to overcome the great mathematical difficulties involved, they imagined, as has already been said, the actual limited crystal replaced by one extended indefinitely. Thus the disturbing effect of the surface on the interior could be eliminated, so that now all atoms were exposed to the same conditions. Here also the main problem is again to determine the elastic spectrum, or—if we dispense with the exact calculation of the proper frequencies—at least to discover the law, according to which the proper (or natural) frequencies are distributed among the different regions of frequency. This problem was first solved by *Born* and *Kármán* for regular crystals. The laws thus obtained were then extended to the case of simple point-lattices of arbitrary symmetry, and finally, *Born* deduced them, in his "Dynamics of the Crystal Lattice," for the most general form of space-lattice.[133]

These most general space-lattices arise from the periodic repetition in space of a definite group of atoms and electrons (basic group) which on the whole is electrically neutral, and is enclosed in a parallelopiped of space, the "elementary parallelopiped." In Fig. 4 such a lattice, in this case, however, plane, is illustrated, in which the basic group consists of three particles ($\cdot$ o $\times$). All $\cdot$ particles form together a simple lattice, as do the o and $\times$ particles. We have in this way three interlocked simple lattices.

Thus, for example, the halogen compounds of the alkalies (NaCl, LiCl, KCl, KBr, KI, RbCl, RbBr, RbI, and so forth) form cubic space-lattices, in which the lattice points are alternately occupied by the positive alkali ion and the negative halogen ion (see Fig. 5). If we regard the whole cube here pictured as the "elementary cube," then the basic group would contain eight particles, namely, four ions of each sort (they are numbered here).   We have thus eight interpenetrating simple lattices.   Every four of them would, however, consist of the same kind of particle.   Hence it is advisable to select in this case in place of the cube the rhombohedron (double-lined in

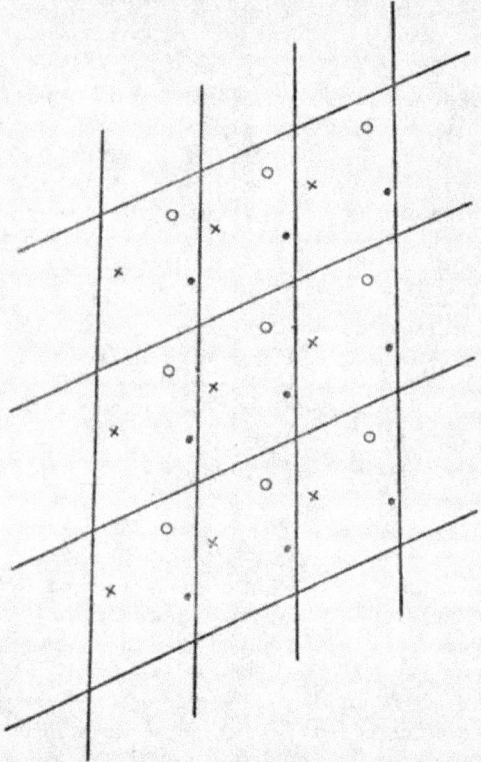

Fig. 4.

the figure) as the elementary parallelopiped.   Then the basic group consists only of the two different particles 1 and 8, of which the one lies in a corner, the other in the middle of the parallelopiped.   In fact we can get the whole lattice by displacing the basic group in the direction of the three rhombohedral edges, a distance equal to a whole multiple of the length of the edge. The lattice consists therefore, according to this view, of two interlaced simple cubical atomic lattices. Furthermore, they are "surface-centred" lattices, that is to say, such that not only the corners of the cubes, but also the middle points of the cube-surfaces, are occupied. If in the most general case the basic group contains s different particles, the lattice consists of s interlaced simple lattices.

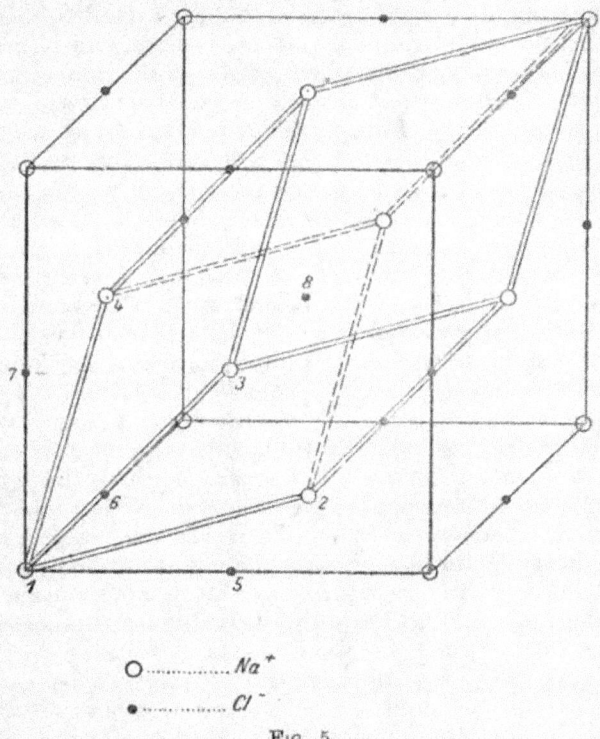

○ .... Na^+

● .... Cl^-

Fig. 5.

In order now to get a general view of the laws which govern the elastic spectrum of such a most general crystal, we proceed according to *Born* and *Kármán* as follows: We imagine an elastic wave of definite wave-length and definite direction (the normal to the wave front) passing through the crystal. For each wave thus defined there are $3s$ natural frequencies with periodicities ν_1 ν_2 ν_3 . . . ν_{3s}. The first three frequencies ν_1, ν_2, ν_3 correspond to those natural frequencies of the crystal, by which the single interpenetrating simple lattices are similarly distorted to a first approximation without being compelled to move relatively to one another. These are the three ordinary acoustic natural periods (one longitudinal, two transverse). The remaining $3(s - 1)$ frequencies, on the other hand, correspond to another type of motion of the crystal, namely, to those natural frequencies with which the single simple lattices oscillate with respect to one another without distortion. If the basic group contains only one particle ($s = 1$), i.e. if the crystal consists of only a simple lattice, this second type of motion disappears altogether, and we are left with only the three acoustic natural frequencies ν_1, ν_2, ν_3. If, on the other hand, we are dealing with a crystal, say of the halogen compounds of an alkali, for example, rock-salt (NaCl), $s = 2$, there exist, as we have seen, besides the three acoustic oscillations, three further natural frequencies of the second type. In consequence of the regular crystal character of the alkaline halides, these three natural frequencies exactly coincide, at any rate for long waves, and give rise to that motion in which the sodium lattice vibrates approximately as a rigid structure against the likewise rigid chlorine lattice. We see at once that it is just the natural frequency last considered that will play the chief part in the optics of these crystals. For when an electromagnetic wave meets the crystal, the sodium ions are driven by the electric force of the wave to the one side, and the oppositely-charged chlorine atoms are drawn to the opposite side. It is thus just the type of vibration described above that is brought about. If the frequency of the external wave approaches closely to that of the natural period, resonance occurs. These infra-red vibrations, therefore, are what determine the course of the refractive index, especially in the infra-red. They are the so-called "infra-red dispersion frequencies." It is also in their neighbourhood that the places of metallic reflection lie which are detected by the method of residual-rays.

What has just been stated for the special case $s = 2$ (alkaline halides) may, of course, be immediately generalised.

For if the basic groups consists of s different particles, it is just the $3(s - 1)$ natural frequencies that determine the dispersion of the crystal. Among them are those, in the neighbourhood of which the regions of metallic reflection (residual rays) lie. If the basic group contains p positive atomic residues and $s - p$ electrons, the frequencies $\nu_4 \ldots \nu_{3s}$ fall correspondingly into two classes: the first class consists of $3(p - 1)$ infra-red frequencies, which arise from the atomic residues; the second consists of $3(s - p)$ ultra-violet frequencies, which are to be ascribed to the influence of the electrons. The infra-red natural frequencies decide the course of the refractive index in the infra-red, the position of the residual rays, and, as we shall see, the atomic heats; the ultra-violet natural frequencies, on the other hand, determine chiefly the refractive indices in the visible and ultra-violet. Incidentally, the general lattice-theory of *Born* [134] confirms the law previously enunciated by *Haber* [135] that the frequencies of the first class (infra-red) bear the same ratio to the second (ultra-violet) class, as regards order of magnitude, as the square root of the mass of the electron bears to the square root of the mass of the atom.

After this digression let us now return to our starting-point. Up to the present we have always considered a wave of definite length λ and with a definite normal direction n, and we have seen that corresponding to it there are, in the most general case, $3s$ natural frequencies $\nu_1 \ldots \nu_{3s}$. Let us now allow the wave-length λ to vary continuously, keeping the wave-direction constant, by going from infinitely long waves to the smallest. Then each of the $3s$ natural frequencies will also vary continuously, and will pass through a continuous range of values. In other words, the $3s$ natural frequencies are certain functions of the wave-length λ:

$$\nu_i = f_i(\lambda).$$

From this, however, we learn the fundamental fact that all these ranges of values of the single natural frequencies *are only finite in extent and that, therefore, each of the $3s$ continua of frequencies automatically breaks off at a highest limiting frequency.* "Automatically," i.e. without our arbitrary assistance (as in *Debye's* case), solely on account of the analytical form of the function f_i. This is explained by the fact that the wave-length λ of possible waves in the crystal has a lower limit set to it : waves of length below a certain lowest value cannot exist. This is most simply recognised from the following instructive example. If we consider a simple cubical lattice having the atomic distance a, and examine, for example, longitudinal waves, which are being propagated along an edge of the cube—so that all atoms on an edge at right angles to this side oscillate in the same phase in the direction of the edge—then we see at once that the smallest wave that is possible here has the length $\lambda_{min} = 2a$. For this wave, namely, successive planes of the cube swing in opposite phase, that is, "against" one another. The functional relation between ν and λ assumes the special form : [136]

$$\nu = \nu_m \sin\left(\frac{\pi a}{\lambda}\right) \qquad . \qquad . \qquad (47)$$

For infinitely long waves ($\lambda = \infty$), $\nu = 0$; if we pass on to shorter waves, ν increases continuously, until, for $\lambda = 2a$, it reaches its maximum value ν_m. At this limiting frequency ν_m the range of possible ν's breaks off automatically.

Up to the present we have given the wave-direction (n, the direction of the normal) a certain fixed value, and have allowed the wave-length λ to vary. We now give the wave-direction by degrees other values, and at each step we allow the wave-length to vary from the value ∞ to the least possible value. Then the nature of the functional dependence of the magnitude ν_i or λ, and the position of the limiting frequencies also change continuously with the wave-direction, so that we may say: the $3s$ natural frequencies are, in general, continuous functions of the wave-length λ and of the wave-direction n :

$$\nu_i = f_i(\lambda, n), \qquad (i = 1, 2, 3, \ldots 3s) \qquad . \qquad (48)$$

In it, each of the functions f_i breaks off automatically for a minimum value of the wave-length at an upper limit $(\nu_i)_{max}$, which itself still depends on the wave-direction. These equations express the *law of dispersion* of waves in crystals, for they determine for each wave the $3s$ frequencies

ν_i and hence also tell us how the rates of propagation $q_i = \nu_i \cdot \lambda$ depend on the wave-length and the wave-direction. The dispersion law becomes particularly simple in the region of long waves: for the three acoustic vibrations the relations [137]

$$\nu_1 = \frac{q_1(n)}{\lambda}, \qquad \nu_2 = \frac{q_2(n)}{\lambda}, \qquad \nu_3 = \frac{q_3(n)}{\lambda} \qquad . \quad (49)$$

hold. In them the three magnitudes $q_1(n)$, $q_2(n)$, and $q_3(n)$ are three, in general different, functions of the wave-direction. And further, these are the three velocities of propagation of the three acoustic vibrations. In the region of long waves, therefore, the three velocities of propagation of the three slow acoustic vibrations are independent of the wave-length to a first approximation.

The dispersion law (for long waves) assumes a very different appearance for the $3(s - 1)$ rapid vibrations $\nu_4, \nu_5 \ldots \nu_{3s}$. It assumes the form

$$\nu_i = \nu_i^0 + \frac{p_i(n)}{\lambda} \qquad (i = 4, 5, \ldots 3s) \qquad . \quad (50)$$

here the ν_i^0's are constants, the $p_i(n)$'s are again certain functions of the wave-direction. The velocities of propagation here assume the values

$$q_i = \nu_i \lambda = \nu_i^0 \lambda + p_i(n) \qquad . \qquad . \qquad . \quad (51)$$

and would thus be linear functions of the wave-length.

We may summarise thus: *the elastic spectrum of the most general crystal, the basic group of which contains s particles, consists of 3s separate parts (" branches "). Each part consists of a finitely extended continuum of frequencies. The three first parts contain the totality of all slow, acoustic natural frequencies (sometimes called " characteristic "). The remaining $3(s - 1)$ parts include the rapid (infra-red and ultra-violet) natural frequencies, which play the chief part in determining the optical dispersion and the positions of metallic reflection.*

§ 8. Continuation. The Law of Distribution of the Natural Frequencies

While this knowledge of the general character of the elastic spectrum is, as we shall soon see, of great value, it is none the less insufficient for the question of the energy-content and molecular heat of the crystal, inasmuch as, even for the simplest crystal, a strict calculation of the elastic spectrum is not possible at the present time. We know, however, on the other hand, that we need not know the whole details of the elastic spectrum to calculate the energy-content and the molecular heat, but that it suffices to know the law according to which the natural frequencies are distributed over the elastic spectrum (or its individual "branches"). This is the more true, the closer together the natural frequencies lie. Now, in reality the finite crystal possesses, if it consists of the basic group (of s particles) N times repeated, $3Ns$ natural frequencies, *which are distributed so that N frequencies fall to each of the $3s$ branches of the spectrum.* If N becomes infinite, the N individual natural frequencies of each branch merge into one another to form a continuum, and we get exactly the elastic spectrum that we have just been considering. We see from this, that the more we are justified in replacing the finite crystal by one of infinite extent the better our results if we know only the distribution law of the natural frequencies (without knowing their position exactly).

The law of distribution of the natural frequencies, which was discovered by *Born* and *Kármán* and extended by *Born* in his "Dynamics of the Crystal Lattice" to the most general type of crystal, may be formulated thus: *Select from the totality of all elastic waves the small group, whose lengths lie between λ and $\lambda + d\lambda$, and whose normal direction lies in the elementary solid angle* [138] $d\Omega$. *Each of the $3s$ branches of the spectrum then contribute* $\dfrac{V}{\lambda^4}\, d\lambda d\Omega$ *natural frequencies to this group.* Here V denotes the volume of the finite crystal.

§9. Continuation. The Atomic Heats at Low, very Low, and High Temperatures

The knowledge of this law of distribution allows us to write down at once the thermal capacity of the crystal consisting of Ns particles. From (42) it is:

4

$$\left.\begin{aligned}
\Gamma_v &= k \sum_{i=1}^{3Ns} \frac{\left(\dfrac{h v_i}{kT}\right)^2 e^{\frac{h v_i}{kT}}}{\left(e^{\frac{h v_i}{kT}} - 1\right)^2} \\
&= kV \sum_{i=1}^{3s} \int_0^{4\pi} d\Omega \int_{\lambda=\lambda_m(n)}^{\lambda=\infty} \frac{d\lambda}{\lambda^4} \frac{\left(\dfrac{h v_i}{kT}\right)^2 e^{\frac{h v_i}{kT}}}{\left(e^{\frac{h v_i}{kT}} - 1\right)^2}
\end{aligned}\right\} \qquad (52)$$

This formula is to be interpreted as follows: the natural frequencies v_i are, by (48), to be expressed as functions of the wave-length λ and the wave-direction n: then the integration is to be performed with respect to λ from the smallest wave-length $\lambda_m(n)$, which itself depends upon the wave-direction n, up to the maximum $\lambda = \infty$. The result of this integration still depends on the wave-direction and the index i. Finally, integration is to be performed over all directions (that is, over all elementary solid angles between 0 and 4π) and summation over all $3s$ branches of the spectrum. But we have seen that the $3s$ branches of the spectrum fall into two groups. The first 3 branches ($i = 1, 2, 3$) contain the totality of slow acoustic natural frequencies; for these branches we have the dispersion law (49) which is valid for long waves. The remaining $3(s - 1)$ branches contain the totality of the quick (infra-red and ultra-violet) natural frequencies, with the entirely different type of dispersion law (50), which also holds for long waves. Hence the suggestion naturally occurs of dividing the sum $\sum_{i=1}^{3s}$ of (52) into two parts, corresponding to the two different groups of frequencies and of writing

$$\Gamma_v = \Gamma_v^{(1)} + \Gamma_v^{(2)}$$

where

$$\left.\Gamma_v^{(1)} = kV \sum_{i=1}^{3} \cdots \; ; \qquad \Gamma_v^{(2)} = kV \sum_{i=4}^{3s} \cdots \right\} \qquad (53)$$

These still very complicated formulæ may, according to *Born*, be brought into a very simple and comprehensive form by limiting our considerations to low temperatures and introducing certain approximations. As we have already recognised, at low temperatures only the long waves contribute to the energy-content. Hence we shall apply in formula (53) all those approximations which are introduced by confining ourselves to long waves. Let us consider first $\Gamma_v^{(2)}$. Here we set in place of the ν_i's of (50) the *constant* values ν_i^0, which are independent of the wave-length λ and of the wave-direction. If we do this, we can place the constant factors

$$\frac{\left(\frac{h\nu_i^0}{kT}\right)^2 \cdot e^{\frac{h\nu_i^0}{kT}}}{\left(e^{\frac{h\nu_i^0}{kT}} - 1\right)^2}$$ in front of both integration signs, and write

$$\Gamma_v^{(2)} = k \sum_{i=4}^{3s} \frac{x_i^2 e^{x_i}}{(e^{x_i} - 1)^2} \cdot \left[V \int_0^{4\pi} d\Omega \int_{\lambda_m(n)}^{\infty} \frac{d\lambda}{\lambda^4} \right], \text{ where } x_i = \frac{h\nu_i^0}{kT}.$$

The factor in square brackets has, however, a simple meaning. From the law of distribution of the natural periods we see, namely, that this factor gives the sum-total of all natural frequencies that occur in one of the $3s$ branches of the spectrum; it therefore has the value N, which as has already been said, is the number of basic groups which go to make up the crystal. If we choose the piece of crystal under consideration such that its size is so that N is equal to the *Avogadro* number, then if we remember that $Nk = R$ for $\Gamma_v^{(2)}$, the expression

$$\Gamma_v^{(2)} = R \cdot \sum_{i=4}^{3} \frac{x_i^2 e^{x_i}}{(e^{x_i} - 1)^2}, \quad \text{where } x_i = \frac{h\nu_i^0}{kT} \quad . \quad (54)$$

follows. If we compare this result with (34) we see that $\Gamma_v^{(2)}$ —excepting for the missing factor 3—consists of $3(s-1)$ *Einstein* functions. We write the expression in the form

$$\Gamma_v^{(2)} = \frac{1}{3} \sum_{i=4}^{3s} E(x_i), \quad \text{where } x_i = \frac{h\nu_i^0}{kT} \quad . \quad (55)$$

in which the abbreviation is obvious. The fact that, in using these approximations, we come across *Einstein* factors, i.e. that we encounter the "monochromatic" theory, might have been anticipated. For since we treated the ν_i's here as *constants* that are quite independent of wave-length and wave-direction, these vibrations represent processes which have nothing to do with the propagation of elastic waves in the crystal as a whole: and this means that the individual particles, uncoupled as it were, perform $3(s-1)$ monochromatic vibrations.

The approximate evaluation of the first part $\Gamma_v^{(1)}$ is quite different. For here we have to use for the frequencies ν_1, ν_2, ν_3, the relations (49), which connect the three acoustic natural frequencies with wave-length and wave-direction. Here we have therefore to deal with three real elastic oscillations, which are propagated in the crystal with the three different acoustic velocities $q_1(n)$, $q_2(n)$, $q_3(n)$, each of which depends on the direction (n). The crystal acts here as a dynamic whole, exactly as in *Debye's* point of view. Hence we may conjecture that $\Gamma_v^{(1)}$ allows itself to be brought into the form of three *Debye* functions (45). The more exact calculation confirms this supposition, and gives us [140]

$$\Gamma_v^{(1)} = 3R \sum_{i=1}^{3} \frac{1}{\bar{x}_i^3} \int_0^{\bar{x}_i} \frac{x^4 e^x dx}{(e^x - 1)^2} \qquad . \qquad (56)$$

which, taking *Debye's* formula (45) into consideration, we may write in the following immediately intelligible form :

$$\Gamma_v^{(1)} = \frac{1}{3} \sum_{i=1}^{3} D(\bar{x}_i) \qquad . \qquad . \qquad (57)$$

The three magnitudes $\bar{x}_i$ here play the part of three upper limits of frequency. Their values are

$$\bar{x}_i = \frac{h \bar{q}_i}{kT} \cdot \sqrt[3]{\frac{3N}{4\pi V}} \qquad . \qquad . \qquad (58)$$

where the three magnitudes $\bar{q}_i$ represent certain mean directions of the acoustic velocities, which therefore no longer depend on the wave-direction. From (55) and (57) we get for the thermal capacity of the piece of crystal considered

$$\Gamma_v = \frac{1}{3}\left\{\sum_{i=1}^{3} D\left(\overline{x_i}\right) + \sum_{i=4}^{3s} E(x_i)\right\} \qquad . \qquad . \quad (59)$$

Now, since N particles of each of the s different kinds of particles are present, that is one gramme-atom of each kind of particle exactly—for N is the *Avogadro* number—the piece of crystal contains s gramme-atoms of different sorts of particles. If, therefore, we cut the crystal into s equal pieces in such a manner, that each piece comprises only $\frac{N}{s}$ basic groups, then each of these pieces contains a so-called "mean" gramme-atom. Hence if we now consider only a *single* one of these pieces, its thermal capacity is $\frac{\Gamma_v}{s}$; we call it the "mean atomic heat" $\overline{C_v}$, and we may write

$$\overline{C_v} = \frac{1}{3s}\left\{\sum_{i=1}^{3} D(\overline{x_i}) + \sum_{i=4}^{3s} E(x_i)\right\} \qquad . \qquad . \quad (60)$$

Here the $\overline{x_i}$'s have the same meaning as in (58). For the piece of crystal now under consideration consists of $\frac{N}{s}$ basic groups, and has therefore the volume $\frac{V}{s}$. Formula (58), however, obviously remains unchanged when we replace in it N and V by $\frac{N}{s}$ and $\frac{V}{s}$. The quantity $\frac{V}{s}$, the volume of a mean gramme-atom, is also called the *mean atomic volume*.

In the case of chemical compounds, in which several sorts of atoms occur in the basic group, and also in the case of polyatomic elements, in which the basic group contains several particles of a like sort, we frequently speak of the *molecular heat*. In doing so, we follow the usual chemical conception, inasmuch as we imagine the s particles of the basic group divided into one or several sub-groups, and regard each sub-group, taken alone, as a molecule. If then the molecule contains q atoms, then $q\bar{C}_v$ is the *mean molecular heat ;* for example, the basic group of rock-salt (NaCl) contains one sodium ion and one chlorine ion. The whole piece of crystal, which, by definition, contains $\dfrac{N}{s} = \dfrac{N}{2}$ basic groups, comprises therefore $\dfrac{N}{2}$ sodium ions and the same number of chlorine ions, that is to say $\dfrac{N}{2}$ " NaCl-molecules." q is in this special case equal to 2. Hence $2\bar{C}_v$ represents the thermal capacity of N " NaCl-molecules," that is, the mean molecular heat of rock-salt.

If among the s particles of the basic group there are p atomic residues and $s - p$ electrons, the number of *Einstein* factors in (59) reduces to $3(p - 1)$, since the $3(s - p)$ ultra-violet frequencies arising from the $s - p$ electrons contribute only in a vanishingly small degree to the atomic heat as compared with the infra-red. We thus arrive at the law : *the mean molecular heat of a crystal whose basic group includes p (similar or different) atomic residues, is made up, at a sufficiently low temperature, of three* Debye *terms (with, in general, three different upper limits of frequency) and* $3(p - 1)$ Einstein *terms (in which the* $3(p - 1)$ *infra-red natural frequencies for long waves appear as frequency numbers).*

When we descend to the *lowest* temperatures, the *Einstein* terms disappear exponentially, and only the three *Debye* terms remain, for these, as we know, decrease much more slowly. In them we can further replace all the upper limits of the three integrals (see (56)) by ∞, so that the integrals thereby become numerical constants. Remembering (58) we get the fundamental law, that *the molecular heat of every crystal at the lowest temperatures is proportional to the third power of the absolute temperature.* So the general lattice theory confirms *Debye's* result. The formula obtained has the following simple form : [141]

$$\bar{C}_v = \frac{16\pi^5 k^4 \bar{V}_A}{5h^3} \cdot \left(\frac{T}{q}\right)^3 \qquad . \qquad . \qquad . \quad (61)$$

where V_A is the " mean atomic volume "

$$\left(= \frac{\text{mean atomic weight}}{\text{mean density}} \right)$$

and $\bar{q}$ represents a quantity which, if suitably defined, may be called the mean acoustic velocity, introduced in place of the three different acoustic velocities $\bar{q}_1$, $\bar{q}_2$, $\bar{q}_3$.

Also in the other extreme case, for high temperatures, a very useful formula can be obtained, as *H. Thirring*[142] showed. He started from (52) and developed the exponential functions in series. The following value is then obtained for the mean atomic heat:

$$\bar{C}_v = 3R\left\{1 - \frac{J_1}{12}\left(\frac{h}{kT}\right)^2 + \frac{J_2}{240}\left(\frac{h}{kT}\right)^4 - \frac{J_3}{6050}\left(\frac{h}{kT}\right)^6 + \ldots\right\}(62)$$

where the coefficients J_1, J_2, J_3, ... depend in a complicated manner on the elastic constants of the crystals, the atomic masses, and the atomic distances.

§ 10. Tests of the Born-Kármán Theory

How do matters stand with regard to the testing of the *Born-Kármán* Theory? We see at once that it is incomparably more difficult than in the case of *Debye's* Theory: for even in simple cases, the calculation of the mean atomic heat of a crystal is very complicated, and requires above all a more exact knowledge of its elastic behaviour than we at present possess. Only by restricting our attention to low and very low temperatures on the one hand, where the formulæ (60) and (61) may be applied, and, on the other, to the region of high temperatures, within the limits of applicability of *Thirring's* formula (62), are we enabled to carry our calculations for a number of simple substances to the point of comparison with experimental results. *Born* and *Kármán* themselves, in one of their first publications[143] tested the formula (61), valid for the lowest temperatures (*Debye's* T^3-law), by comparing its results with those of experiment. They limited themselves in this case to metals (Al, Cu, Ag, Pb) which, however—at any rate in the usual form—are not proper crystals, but irregular crystalline aggregates. For this reason, they proceeded as if the metal were an *isotropic* body, and obtained the mean acoustic velocity—the only quantity in (61) which in general requires extensive calculation—from the following relation which holds for isotropic bodies : [144]

$$\frac{3}{\bar{q}^3} = \frac{1}{q_l^3} + \frac{2}{q_t^3} . \qquad . \qquad . \qquad . \qquad (63)$$

Here q_l and q_t are the velocities of propagation of the longitudinal and transverse elastic waves, magnitudes, therefore, which may be simply calculated from the two elastic constants of the isotropic body and its density.[145] The agreement of the values of C_v thus found with the experimental data is, especially in the case of Al and Cu (and also Pb), quite good. A. Eucken [146] has, however, pointed out rightly, that no weight should be attached to this agreement. For the values of the elastic constants which Born and Kármán used for calculating q_l and q_t are those which are correct at the ordinary room temperature. If we take their dependence on temperature into account, the good agreement between theory and experiment disappears. Metals are, indeed, not isotropic bodies, and hence it is not permissible to use the observable elastic constants, which depend upon temperature, in calculating $\bar{q}$.

Matters are much more favourable in the case of real crystals, in which, as experiments by E. Madelung [147] show, the elastic constants vary very little with temperature. But here the calculation of the mean acoustic velocity $\bar{q}$ gives rise in general to notable difficulties,[148] which may, however, be cleared away in simple cases by a very practical method due to L. Hopf and G. Lechner.[149] Hopf and Lechner were thus enabled successfully to carry out the calculations for sylvin (KCl), rock-salt (NaCl) fluor-spar (CaF$_2$) and pyrites (FeS$_2$). They proceeded to calculate the quantity $\bar{q}$ from the observed values of $\bar{C}_v$, assuming the correctness of formula (61), and they then compared these with the value of $\bar{q}$ calculated from elastic data. The result showed very satisfactory agreement.[150]

It is of particular interest to test the very clear formula (60) which gives the mean atomic heat as a sum of three Debye functions and $3(s - 1)$ Einstein functions. Here the three infra-red natural frequencies $\nu_4^0, \nu_5^0, \nu_6^0$ coincide, and the three *Einstein* functions become equal to one another. If we introduce the further approximation of replacing the three different quantities $\bar{x}_i$ in the *Debye* formula by a mean value $\bar{x}$, it follows that

$$\bar{C}_v = \tfrac{1}{2}\{D(\bar{x}) + E(x)\} \quad . \quad . \quad . \quad (64)$$

In this we use the value of $\bar{x}$ deduced from formula (58) by merely replacing $\bar{q}_i$ in it by a mean value $\bar{q}$, which can be calculated by the method of *Hopf* and *Lechner* just mentioned. x, on the other hand, according to (54), $= \dfrac{h\nu^0}{kT}$, where ν^0 is the infra-red natural frequency of the crystal (for long waves), which may be determined from the dispersion in the infrared or by the method of residual rays.

Formula (64) had already been given, previously to *Born*, by *W. Nernst*,[151] who, however, based his argument on a supposition which is no longer tenable. *Nernst* started from the conception that, for example, in the case of rocksalt, the NaCl-molecules are located upon the points of the space-lattice, and that the most general state of oscillation of the lattice arises from the superposition of two modes of motion, firstly the oscillation of the whole molecules in the lattice-structure, which give a *Debye* term, and secondly the intra-molecular oscillations of the two atoms, which, being almost monochromatic, lead to an *Einstein* term. The agreement of the *Born-Nernst* formula (64) with the experimental data is not very satisfactory in the case of NaCl and KCl, but much better in the case of AgCl, which belongs to the same crystal type.[152] The reason for this is believed by *E. Schrödinger*[153] to lie in the excessively rough approximation inherent in formula (64).

Finally, *Thirring's* formula (62) has also been tested, by *Thirring* himself,[154] for NaCl, KCl, and, by neglecting certain factors, for CaF_2 and FeS_2. Taking into account the variation of the elastic constants with temperature (which, however, is to be regarded as uncertain and provisional since the values are only obtained by interpolation) he found good agreement between theory and experiment. In connection with the *Thirring* formula, *Born*[155] has also calculated the atomic heat of diamond and compared it with experiment. Since in this case, however, the elastic constants were unknown, *Born* proceeded to evaluate the curves of atomic heat for various possible values, and to select from them that curve which conformed most closely to the results of observations. Thus, for example, the value $0.63 \times 10^{-12} \left[\dfrac{cm^2}{dyne} \right]$ was obtained for the compressibility; this is in satisfactory agreement with the value, probably too small, measured by *W. Richards*, viz. $0.5 \times 10^{-12} \left[\dfrac{cm^2}{dyne} \right]$.

From all this we see that the possibilities of testing the *Born-Kármán* Theory of Atomic Heats, partly on account of the great difficulties of calculation, partly on account of our insufficient knowledge of the elastic behaviour of crystals, are exceedingly sparse, so that for the present *Debye's* much more tractable formula (if necessary, with the addition of *Einstein* terms) appears more useful. If, in spite of this fact, so much space has been devoted here to the *Born-Kármán* Theory, the reason is to be sought in the conviction that this theory has gone much further than that of *Debye* into the kernel of the matter. For, without a more exact treatment of the structure of the space-lattice and its dynamics, our knowledge of the nature of the solid state must without doubt remain faulty.

§ 11. The Equation of State of a Solid Body

Linking up with this new development of the theory of atomic heats, a number of investigators, chiefly *E. Grüneisen*,[156] *S. Ratnowski*,[157] and *P. Debye*,[158] have worked out a theory of the solid state with the object of creating as a counterpart to the Kinetic Theory of Gases a Kinetic Theory of Solids. One of the main problems in this connexion is to formulate an " Equation of State," that is, a relation between pressure (p), volume (V), and temperature (T), a problem, which, according to the doctrine of thermodynamics, is to be regarded as solved as soon as the "free energy" F of the body is known as a function of the temperature and the volume.[159] Then the pressure, for example, will follow from the simple equation

$$p = - \left(\frac{\partial F}{\partial V} \right)_T \qquad \qquad . \qquad . \qquad . \qquad (65)$$

which, as a relation between p, V, and T, gives the equation of state at once. If this is known, we have mastered quantitatively the behaviour of the body for all changes of state. For example, the coefficients of expansion a, and the compressibility κ, result from the well-known formulæ

$$a = \frac{1}{V_0}\left(\frac{\partial V}{\partial T}\right)_p ; \; \kappa = -\frac{1}{V}\left(\frac{\partial V}{\partial p}\right)_T \qquad . \qquad . \; (66)$$

(V_0 is the volume at the zero-point.)

P. *Debye* [160] was the first to draw attention to the fact that the model of the solid body which forms the basis of the atomic heat theories of *Einstein, Debye,* and *Born-Kármán,* is necessarily too highly idealised ; for *this idealised solid body has,* as is easily seen, *a zero coefficient of expansion.* In fact, if, as has always been assumed hitherto, the forces which pull the atoms back into their position of equilibrium are proportional to the first power of their relative displacements (assumption of quasi-elasticity, *Hooke's* Law), then the atoms will execute *symmetrical* oscillations about this position of rest. If this supposition, viz. *Hooke's* Law, be valid for all temperatures, then the mean volume of the body—that is, the volume that it possesses when all atoms are exactly in their positions of rest—must be just as often overshot as undershot, however great the amplitude of the heat-vibrations may be. Hence, if we warm the body from zero until it possesses the volume V_0, and if we assume that all atoms are at rest at zero, then its mean volume at any temperature will also be equal to V_0. The body, therefore, does not change its mean, observable volume with rise of temperature ; its coefficient of expansion is therefore 0. If we desire to represent the actual behaviour of the solid body, namely, its expansion when heated, as known to us from thousandfold experience, we are necessarily obliged, according to *Debye,* to replace *Hooke's* Law of Force by an expression involving higher powers of the variation of atomic distance. Then the oscillations of the atoms become unsymmetrical, and there occurs a displacement of their position of rest as the energy of vibration increases. If we arrange the generalisation of *Hooke's* Law so that a greater force is necessary to bring the atoms nearer together than to separate them, then the change in the position of rest occurs in such a manner that for increasing energy of vibration, that is, for rise of temperature, the relative distances of the atoms increase, and hence the body increases in volume. *Debye* has extended the theory in this sense. Among other things this gives us the law previously deduced by *Grüneisen*[161] that at sufficiently low temperatures the thermal coefficient of expansion a is proportional to the specific heat. Moreover, the very small change in compressibility with temperature is well accounted for on *Debye's* Theory.

§ 12. The Thermal Conductivity of Solid Bodies according to Debye

The importance of *Debye's* Theory is by no means confined to thermal expansion. On the contrary, it became manifest that another important group of phenomena require this generalisation of *Hooke's* Law. In the idealised solid body, in which the elastic forces obey *Hooke's* Law, the elastic waves will become superposed without disturbance, and will penetrate the whole body without becoming weakened. If we imagine the idealised body as a horizontal, infinitely extended plate of finite thickness, and if we transmit a powerful motion (high temperature) to the upper layer of atoms, while we keep the lower layer at rest (i.e. at zero temperature), then an elastic energy current (heat current) will pass continually from above to below. An energy gradient (temperature gradient) does not, however, exist in the body, since, on account of the undamped character of the wave, the mean density of energy is everywhere the same. Since, in general, the conductivity for heat is equal to the flux of heat divided by the gradient of temperature, it follows that *the idealised solid body possesses an infinite thermal conductivity*. The case becomes different, however, if we extend *Hooke's* Law in the manner described, and thus pass over to the "real" solid body. The waves in the body will then, on account of the departure of the equations of motion from linearity, no longer pass over one another undisturbed. On the contrary, an oscillation already present will, in consequence of the fluctuations in density caused by it, disturb the oscillations superimposed upon it, with the effect that a scattering, and therefore a weakening of the waves in the body results, in precisely the same way as a

" cloudy" medium scatters and weakens light passing through it. Hence, in the case taken, a temperature gradient is set up in the plate from the top to the bottom. In the case of the real body we thus arrive at a finite thermal conductivity. The mathematical development of this conception led *Debye* to the law [162] *that the thermal conductivity of crystals is inversely proportional to the absolute temperature* (if we confine ourselves to temperatures which are so high that classical statistics are applicable). This deduction seems to be in excellent agreement with experimental results obtained by *A. Eucken*.[163]

§ 13. The Electron Theory of Metals and its Modification by the Quantum Theory

If matters are already complicated in the intrinsically clear case of crystals, the position becomes still more difficult when we turn to metals which, in general, consist of an irregular conglomerate of crystallites. In this case the conductivities, namely, of heat and electricity, are particularly deceptive. According to the classical theories of *P. Drude*,[164] *E. Riecke*,[165] and *H. A. Lorentz*,[166] these phenomena are brought about by the free conductivity-electrons, which, like gas-molecules, fly about in the space between the fixed atomic residues, exchange energy with these upon collision, and so take part in the establishment of thermal equilibrium. Thus the conduction of electricity is explained as follows: in a piece of metal of uniform temperature, an equal number of electrons fly, on the average, in each direction through an element of surface. Hence, on the average, there is no transport of electrical charges through this element of surface, that is, no electric current is flowing in the piece of metal. If now we apply a potential difference to the ends of the metal, an electric field exists in the metal, and this field impresses upon the electrons during their "free paths" (i.e. their paths between two encounters with atoms) a certain *one-sided* additional velocity which is superimposed upon the irregular heat-motion. Now, therefore, more electrons will pass per second through the element of surface in one direction than in the other, and since the electrons carry a negative charge, and so move against the field, i.e. in a direction opposite to the field, we have now an *electric current* in the metal. The mathematical calculation of this simple conception gives for the electrical conductivity σ of the metal [167]

$$\sigma = \frac{Ne^2l}{2mq} \qquad . \qquad . \qquad . \qquad . \qquad (67)$$

Here N is the number of electrons per unit of volume, e and m charge and mass of the electrons, q their average velocity, and l their free path. If we write the expression (67) in the form

$$\sigma = \frac{Ne^2lq}{4 \cdot \frac{1}{2}mq^2} \qquad . \qquad . \qquad . \qquad (67a)$$

we may, according to the assumptions of the classical theory, replace the mean kinetic energy $\frac{1}{2}mq^2$ of the electrons by $\frac{3}{2}kT$. For since, as we assumed, the electrons take part in establishing heat-equilibruim, the law of equipartition of kinetic energy applies to their motion, and there is thus allocated to each of the three degrees of freedom of the electrons the energy $\frac{1}{2}kT$. In this way we arrive at the formula

$$\sigma = \frac{Ne^2lq}{6kT} \qquad . \qquad . \qquad . \qquad (67b)$$

Analogously, we get from *Drude's* Theory the coefficient of thermal conductivity [168]

$$\gamma = \frac{1}{2}Nklq \qquad . \qquad . \qquad . \qquad (68)$$

so that a combination of the two formulæ leads to the fundamental relation

$$\frac{\gamma}{\sigma} = \frac{3k^2}{e^2} \cdot T \qquad . \qquad . \qquad . \qquad (69)$$

which is the law of *Wiedemann-Franz* and of *Lorenz*.[169] It states that *the ratio of the thermal to the electrical conductivity has the same value for all pure metals and is proportional to the absolute temperature.*

Thus all appeared in the best of order. The classical theory appeared here also to have worked successfully and the law of equipartition celebrated a triumph. But upon closer inspection, gaps appeared in the apparently solid theoretical structure, and serious doubts arose. For if the free electrons really took part in the thermal equilibrium, and hence claimed their full share, $\frac{3}{2}NkT$ (per unit of volume), in the equal division of kinetic energy, then this share of energy should be plainly noticeable in the atomic heat of the body, namely, to the extent of $\frac{3}{2}N^*k$, where N^* denotes the number of electrons in a gramme-atom. Such an increase in the atomic heat of the metals as compared with the non-metals (which contain no, or vanishingly few, free electrons) has never been observed. This difficulty could be avoided by assuming that the number of electrons is small compared with the number of atoms per unit volume, and then their contribution to the atomic heat would be relatively small. But then we should expect from (67b) much smaller conductivities than experiment has disclosed, unless we were to assume high values, that are improbable, for the mean free path.[170]

Further, *H. A. Lorentz* [171] has shown, as we have seen, that, if the law of equipartition for the motion of the electrons is assumed, the metals would radiate in the region of long waves according to *Rayleigh's* Law, whereas we have unquestionably to expect, especially at low temperatures, the radiation to take place according to *Planck's* Law.

The calculated dependence of the conductivity on temperature can only be made to agree with experience by making particular assumptions at high temperatures, whereas no assumptions seem to be able to make calculation and observation agree for low temperatures. At high temperatures the resistance of the metals increases proportionally to the temperature, that is, σ decreases with $\frac{1}{T}$. This can only be reconciled with (67b), if the product Nlq is independent of the temperature. If we assume with *J. J. Thomson* [172] that N increases proportionately to $\sqrt{T}$, then, since q is likewise proportional to $\sqrt{T}$, l must decrease with $\frac{1}{T}$, a hypothesis which, as we shall see, has latterly been upheld by several investigators.

Now, although the agreement between theory and experiment could thus be compelled by special assumptions at high temperatures, the region of low temperatures revealed itself

as the vulnerable point of the theory. For experiments by
H. Kamerlingh-Onnes [173] in the laboratory for low temperatures at Leyden had shown that the resistance of metals
at very low temperatures (the experiments extended as far
down as 1·6° abs.) falls away to a quite extraordinary degree,
and practically disappears before the zero-point is reached.
At any rate, the resistance cannot, as follows in view of what
has just been said from formula (67*b*), sink proportionately
to only the *first* power of the temperature; on the contrary
the fall is without doubt proportional to a higher power.
That the *Wiedemann-Franz* Law also ceases to be valid in
this region, has been proved by experiments of *C. H. Lees* [174]
and *W. Meissner*.[175]

In order to escape from all these difficulties the quantum
theory was appealed to, and attempts were made, in the most
varied ways, to make it harmonise with the existing theory.
A first attack was ventured by *W. Nernst* [176] and *Kamerlingh-
Onnes*,[177] who gave for the resistance of the metals empirical
formulæ which linked up directly with the form of *Planck's*
energy equation (9) and which gave the change in the resistance with temperature satisfactorily. *F. A. Lindemann* [178]
and *W. Wien* [179] conceived more detailed theories. *Lindemann* accepts in his first paper *J. J. Thomson's* hypothesis,
according to which N is proportional to $\sqrt{T}$, and retains the
equipartition law for the motion of the electrons, so that q
also becomes proportional to $\sqrt{T}$. *Then, according to* (67*b*),
the variation of the resistance $\dfrac{1}{\sigma}$ *with the temperature depends
entirely on the mean free path l.* But this is, according to
well-known considerations of the theory of gases, the greater
the smaller the "radius of action" of the metallic atoms; for
the electrons can pursue greater paths freely, i.e. without
collisions, the smaller the hindrances set in their path. The
novel part of *Lindemann's* Theory is the fact that he brings
the radius of action of the atom into relation with its amplitude of swing in its heat-motion. For it is at once obvious
that the atoms in this heat-motion will cover a greater space
in a given time, and their sphere of action will be the greater,
the larger their amplitude of oscillation, i.e. the higher the
temperature. Thus the mean free path also becomes a

function of the temperature, inasmuch as it is brought into relation with the energy of vibration of the atoms. But, for the latter term, *Lindemann* inserts the value given by the quantum theory, and finds for the resistance the formula [180]

$$W = \frac{A^2}{e^{\frac{h\nu}{kT}} - 1} + \frac{2AB}{\sqrt{e^{\frac{h\nu}{kT}} - 1}} + B^2 \qquad . \qquad (70)$$

where ν denotes the frequency of the atoms (again the monochromatic theory); A and B are constants. For high temperatures W then becomes proportional to the temperature T; for low temperatures W decreases exponentially with $e^{-\frac{h\nu}{2kT}}$ to a constant value B^2. With the help of this formula, *Lindemann* succeeds in representing the observations quite well (the formula contains two constants which can be manipulated); but, since the law of equipartition has been retained for the electrons, the difficulties of the excessive atomic heat and of *Rayleigh's* radiation formula remain. Moreover, this theory is unable to explain the departures from the *Wiedemann-Franz* Law at low temperatures; for the mean free path l—the only quantity dependent on T which occurs in σ —disappears entirely from the formula (69).

W. *Wien* attacked the question much more radically than *Lindemann*. In order once and for all to get rid of the contribution of the electrons to the atomic heat—this is the weak point of all theories which make use of the law of equipartition—he assumed that the electrons do not take part in the thermal equilibrium, but possess a velocity q which is independent of the temperature. Moreover, he makes the number N of the electrons per unit volume equal for all temperatures. Then, according to (67), the variation of $\frac{1}{\sigma}$ with temperature is again determined only by the dependence of the mean free path l on the temperature. *Wien*, in a manner similar to that of *Lindemann*, connects l with the energy of vibration of the metallic atoms, taking, however, the complete elastic spectrum into account according to *Debye*. He thus gets for the resistance the value

5

$$W = \text{const.} \int_0^{\nu_m} \frac{\nu d\nu}{e^{\frac{h\nu}{kT}} - 1} \qquad . \qquad . \qquad . \qquad (71)$$

For high temperatures this formula gives $W = \text{const. } T$, i.e. proportionality with the temperature. For low temperatures it follows that $W = \text{const. } T^2$, i.e. a parabolic decrease. The observations are very well represented by *Wien's* formula. But, above all, the unsatisfactory fact remains that this method does not lead us on to a theory of heat-conduction, unless we make new assumptions, nor to the *Wiedemann-Franz* Law. For, by the condition that the motion of the electrons takes place quite independently of the temperature, *Wien* has taken away the possibility of also ascribing the transport of heat to the electrons.

This difficulty arises again in a more recent paper of *F. A. Lindemann* [181] in which, in continuation of the conceptions of *Born* and *Kármán*, the supposition is introduced that—just as the atoms in a crystal—*the electrons in a metal form a lattice*. *F. Haber* [181a] has also adopted a similar hypothesis. The conduction of electricity is then explained by supposing this electron lattice to move practically as a rigid structure relatively to the atomic lattice and so through the metal. This model has many advantages. Since, in the heat-motion, in which the electron lattice naturally takes part, the electronic vibrations, on account of their mass, are extremely rapid (high frequency), these vibrations of the electron, according to *Planck's* formula for the energy, make no appreciable contribution to the atomic heat. In addition the abnormal conductivity (supra-conductivity) which has been observed at very low temperatures may, if we use earlier considerations by *J. Stark,*[182] be explained without difficulty by the conception that at these very low temperatures at which the atomic space-lattice is practically at rest, the electronic lattice glides almost unimpeded through the gaps of the atomic lattice.

G. Borelius,[183] in a sketch which was recently published, uses ideas similar to those of *Lindemann.*

Finally, we may refer to a paper by *K. Herzfeld* [184] which, in contrast to the preceding investigations, attacks the question from a more phenomenological point of view without making use of a particular model. For if, in the formulæ for σ and γ, (67) and (68), we bring into evidence the energy $E = \frac{1}{2}mq^2$ of the electrons by writing the equations thus :[185]

$$\left. \begin{aligned} \sigma &= \frac{Ne^2l}{2\sqrt{2mE}} \\ \gamma &= \tfrac{1}{3}Nl\sqrt{\frac{2E}{m}}\frac{dE}{dT} \end{aligned} \right\} \qquad . \qquad . \qquad . \quad (72)$$

we get

$$\frac{\gamma}{\sigma} = \frac{4}{3e^2} \cdot E\frac{dE}{dT} \qquad . \qquad . \qquad . \qquad . \quad (73)$$

as the expression which represents the *Wiedemann-Franz* Law. *Herzfeld* then shows that, if we compare the result with observation, the formula (73) can be made to agree well with the actual measurements if we set *Planck's* expression $\frac{1}{2}\dfrac{h\nu}{e^{\frac{h\nu}{kT}}-1}$ for E. (The factor $\frac{1}{2}$ has been introduced because the energy of the electrons is solely kinetic.) The values for ν which have to be used stand in no recognisable relation to the atomic frequencies. A paper by *F. v. Hauer* [186] works along similar lines.

If we survey the whole field of the conduction of heat and electricity in metals we recognise that here the last word has not been spoken, and that a great deal of hard work will be necessary to clear up finally the extraordinarily complicated relationships. But much would doubtless be gained for the theory if in future the observations, as far as possible, are no longer made on crystal aggregates, but on metal crystals that are pure and homogeneous.

CHAPTER V

The Intrusion of Quanta into the Theory of Gases

§ 1. The Heat of Rotation of Diatomic Gases according to the Quantum Theory

WHILE the molecular theory of the solid state thus gained new nourishment from the doctrine of quanta, the kinetic theory of gases could no longer be preserved from the influx of the new views. *W. Nernst* [187] had pointed out quite early that quantum-effects are to be expected in the rotation of di- and polyatomic gas-molecules, and also in the oscillation of atoms in the molecule. Let us take as an example the diatomic gas hydrogen, the molecule of which we may picture provisionally as a rigid "dumb-bell" (Fig. 6). The knobs of the dumb-bell are the hydrogen atoms, the grip represents their chemical bond. Such a molecule is known to possess, besides its translatory motion (three degrees of freedom), the possibility of rotating about an axis at right angles to the line joining the atoms (two degrees of freedom, corresponding to the two axes dotted in the figure). Rotation about the line joining the atoms does not—if we accept *Boltzmann's* conception of the absolutely rigid smooth atom—come into play in the exchange of energy by collision and hence in the distribution of energy among the separate degrees of freedom : for this

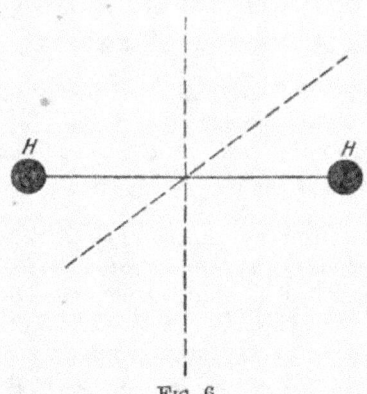

FIG. 6.

rotation cannot be changed by collision. Considered from the new point of view of quanta, which rejects " rigidity " and " smoothness " as an unjustified idealisation, the position is as follows : The moment of inertia of the molecule relatively to the line joining the atoms is extremely small compared with the moment of inertia about either of the axes at right angles to this line. But it is known that rotations which take place about axes with small moments of inertia occur with much greater rapidity than those about axes with large moments of inertia (the same energy having been imparted in each case). If, therefore, we identify the revolutions per second with frequencies, and use the *Planck* energy expression for the energy of rotation (which is not strictly correct quantitatively), a line of argument which has already been frequently applied shows us *that the rotation about the line joining the atoms possesses only a vanishingly small share of the energy.* For the same reason (high frequencies) the degrees of freedom, which correspond to the vibration of the atoms in the molecule, become of importance only at high temperatures. As a result of all this, classical statistics gives us the value $2\dfrac{kT}{2} = kT$

for the mean energy of rotation of the hydrogen molecule; per gramme-molecule it therefore becomes $NkT = RT$.

Hence that part of the molecular heat which arises from rotation is equal to R, that is, about $1\cdot98\ \dfrac{\text{cal.}}{\text{deg.}}$, *and it is independent of the temperature.* In crass contradiction to this, A. Eucken [188] found experimentally that the rotation part of the molecular heat of hydrogen has the value R demanded by the classical theory only at high temperatures. On the other hand, it gradually decreases as we pass to lower temperatures, and approaches asymptotically the value zero at the absolute zero. *In the immediate neighbourhood of absolute zero, hydrogen behaves as a monatomic gas.* Eucken's result was confirmed by experiments conducted by K. Scheel and W. Heuse,[189] who, however, measured the values of the molecular heat only for three temperatures (92°, 197°, and 289° on the absolute scale). This falling off of the rotational heat is without doubt a quantum-effect, similar to the decrease in the atomic heat of solid bodies.

The first attempt to calculate this phenomenon theoretically, is due to *A. Einstein* and *O. Stern*,[190] who proceeded as follows : If J and v are the moment of inertia and the number of revolutions per second of the molecule respectively, then its rotational energy is

$$E_r = \frac{J}{2} \cdot (2\pi v)^2 \qquad . \qquad . \qquad . \qquad (74)$$

If, to simplify matters, we now suppose that all molecules rotate with the same mean number of revolutions $\bar{v}$ per second, then we can introduce for the corresponding mean energy of rotation

$$\bar{E}_r = \frac{J}{2}(2\pi\bar{v})^2 \qquad . \qquad . \qquad . \qquad (75)$$

the theoretical quantum value [191]

$$\bar{E}_r = \frac{h\bar{v}}{e^{\frac{h\bar{v}}{kT}} - 1} \quad \text{(according to } Planck's \text{ first theory)} \qquad (76)$$

or

$$\bar{E}_r = \frac{h\bar{v}}{e^{\frac{h\bar{v}}{kT}} - 1} + \frac{h\bar{v}}{2} \quad (Planck's \text{ second theory)} \quad . \quad (77)$$

From (76) or (77), by combining with (75), we obtain $\bar{v}$ as a function of T. If finally we form $\frac{d\bar{E}}{dT}$, and multiply by the *Avogadro* number N, then we get the share of the energy of rotation in the molecular heat, and we see how it depends on the temperature. It thus appeared that only by using the expression (77) for $\bar{E}_r$ would we be enabled to get a satisfactory connexion agreeing with *Eucken's* measurements, a fact which *Einstein* and *Stern* used at the time as an argument for the existence of a zero-point energy. It must, however, be emphasised that this theory can only be regarded as a first attempt to find general bearings and that it does not fulfil more rigorous requirements. For the *Planck* energy formulæ used, (76) or (77) are valid, as is shown by their genesis, only for configurations whose frequency v is a constant quantity independent of the temperature. Here, on the contrary, we have made use of a mean speed of rotation $\bar{v}$, *dependent on the temperature.*

P. Ehrenfest [192] in 1913 built up a theory of the heat due to rotation on a stricter basis. He had, however, to confine himself to configurations with *one* degree of freedom, that is, to rotations of the molecule around a *fixed* axis, as at that time an extension of the quantum theory to several degrees of freedom had not yet been worked out. The expression thus obtained for the heat of rotation was then, in order to take into account both degrees of freedom, simply multiplied by 2, a method which readily suggests itself, but is not justifiable. *Ehrenfest* started, in his calculation, from the original form of the quantum hypothesis, according to which the energy of linear oscillators may only be whole multiples of $h\nu$, and accordingly made the condition that the rotational energy of a configuration with one degree of freedom (fixed axis) may only consist of whole multiples of $\frac{h\nu}{2}$. The factor $\frac{1}{2}$ appears, because the energy of rotation—in contrast with the vibrational energy of the oscillator—is solely kinetic by nature. The *Ehrenfest* condition is, therefore, according to (74):

$$E_r = \frac{J}{2} (2\pi\nu)^2 = n \cdot \frac{h\nu}{2} \quad (n = 0, 1, 2, 3 \ldots) \quad . \quad (78)$$

hence

$$\nu_n = \frac{nh}{4\pi^2 J} \quad (n = 0, 1, 2, 3 \ldots) \quad . \quad (79)$$

and by substitution in (78)

$$E_r^{(n)} = \frac{n^2 h^2}{8\pi^2 J} \quad (n = 0, 1, 2, 3 \ldots) \quad . \quad (80)$$

Hence the molecules can only rotate with quite definite, discrete speeds ν_n, and correspondingly acquire only a series of discrete rotational energies $E_r^{(n)}$, quite in agreement with the sense of Planck's quantum theory. It is noteworthy that these discrete rotational energies are related to one another as the *squares* of the whole numbers, whereas the energies of the *Planck* oscillators are proportional to the whole numbers themselves. With the discovery of the discrete values (80) of the energy, the *dynamical* part of the problem was solved.

But we require the *mean* energy $\overline{E}_r$ of a totality of N similar molecules. It is here, then, that the second, *statistical* part of the calculation begins. If w_n denotes the probability that a molecule possesses the rotational energy $E_r^{(n)}$ at the temperature T (w_n is therefore the "distribution-function" which has been extended in accordance with the quantum theory), then the mean rotational energy of a molecule is known to be equal

to $\displaystyle\sum_{n=0}^{\infty} E_r^{(n)} \cdot w_n$. Multiplication by N and differentiation with respect to T give us immediately the heat due to rotation.[193] *Ehrenfest* thus obtained for the relationship between the rotational heat and the temperature a curve which could, it is true, be made to agree well with the measurements obtained at low temperatures by choosing the arbitrary constant J (the moment of inertia) suitably, but at high temperatures it showed, before reaching the classical value R, a maximum and a subsequent minimum, which did not correspond with the existing observations.

We may note here an important consequence of equation (79), since it has played a noteworthy rôle in the further development of the quantum theory. If, namely, we write down the angular momentum (the moment of momentum [194]) of the molecule, that is, the quantity $p = J \times 2\pi\nu$, then it follows from (79) that only the special quantum values

$$p_n = \frac{nh}{2\pi} \quad (n = 0, 1, 2, 3 \ldots) . \quad . \quad (81)$$

of the turning-moment exist. This relation may also be deduced directly from the theory of the quantum of action as formulated in (30). For, if we select as our general coordinate, in this case q, the angle of rotation ϕ, then the corresponding momentum or impulse p is known to be none other than the moment of momentum.[195] It follows from this, since p is independent of ϕ, that

$$\int_0^{2\pi} p_\phi d\phi = 2\pi p_\phi = nh \quad . \quad . \quad . \quad (82)$$

in agreement with (81).

In the same way, on the basis of *Planck's* first theory, namely, the conception that the special quantum rates of rotation ν_n are the only possible ones, and using the dumb-bell model, the author [196] has recently carried out the strict calculation for structures with two degrees of freedom (free axes of rotation), making use of the later ideas of the quantum theory. This stricter method likewise gives us curves for the rotational heat which are useless, for they also have a maximum and a subsequent minimum, as in *Ehrenfest's* case. Only by making special subsidiary assumptions, such as excluding certain quantum states, can we get curves which rise steadily with increasing temperature, and which agree, at least to a certain extent, with observation.[197]

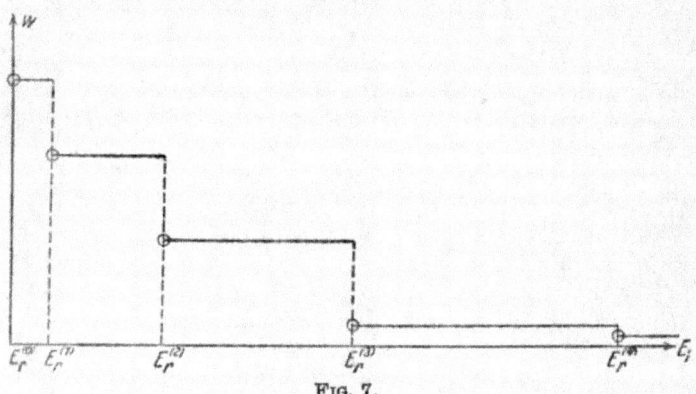

FIG. 7.

Not much more satisfactory results were obtained in those investigations which, again with the use of the dumb-bell model, were based on *Planck's* second theory. According to this theory, the discrete values ν_n of the rotational speeds are not the only possible ones; on the contrary, the molecule can rotate with all rotational speeds between 0 and ∞, and hence can assume all values of rotational energy between 0 and ∞, exactly like the *Planck* oscillators in *Planck's* second theory. The peculiarity of the special quantum values (80) for the energy here consists in the following : imagine the energies E_r plotted as abscissæ (Fig. 7) and the corresponding probabilities w as ordinates; then a *step-ladder* curve results, the steps of which lie exactly at the values $E_r^{(n)}$. The probability that a given value E_r of the rotational energy will appear is therefore *constant within the range of energy between* $E_r^{(n)}$ *and* $E_r^{(n+1)}$, *but changes suddenly at the ends of this range.* According to *Planck's* first theory, which allows only the quantum values $E_r^{(n)}$, the encircled points alone have a meaning. Only at those points is the probability other than zero, while all intermediate values of the energy possess the probability zero, that is, do not occur.

In this case, too, the problem was first solved for one degree of freedom (fixed axis of rotation). *E. Holm*[198] and *J. v. Weyssenhoff*[199] found, in agreement with one another, a steadily rising curve for the rotational heat, which fitted the observations well at low temperatures, but undoubtedly went too high at higher temperatures (from about 140° abs. upwards).

But when the modern development of the quantum hypothesis for several degrees of freedom, to which we shall be introduced later, was available, a stricter calculation for free axes of rotation, i.e. for two degrees of freedom, could be carried out. This problem was attacked on the one hand by *M. Planck*,[200] on the other by *Frau S. Rotszayn*,[201] but was treated differently in each case. *Planck* started with the premise that this problem belongs to the category of so-called "degenerate" problems. This term is to convey the following: the molecule rotates, when no external forces act on it, according to the doctrines of mechanics, with constant speed in a *spatially-fixed* plane. The position of this plane in space must, so *Planck* argues, be of no importance for the statistical state of the molecule. Hence the condition of rotation of the molecule in the sense of the quantum theory is determined by a *single* quantity, namely, the rotational energy. In spite of the fact, therefore, that the problem is originally and naturally a problem of two degrees of freedom—for the position of the molecule in space is determined by two angles—we must, according to *Planck*, treat it in the quantum theory as a problem of only one degree of freedom. The two degrees of freedom coalesce, as it were; they are "*coherent.*"

In contrast to this, *Frau Rotszayn* proceeds to turn the

problem into a non-degenerate one by the addition of an external field, and after solving this problem, reduces the field of force till it vanishes. This method which was also used by the author in the paper above cited, appears to be particularly advantageous, when the calculation is based on *Planck's* first theory, for peculiar difficulties arise in "degenerate" cases. Success here decides in favour of the second method. For while *Planck* finds a curve [202] which rises above the classical value to a maximum, and then *descends* asymptotically towards the value R—and is therefore of no use— the calculation of *Frau Rotszayn* gives a steadily rising curve, which agrees well with the measurements for lower and higher temperatures; only the value observed at $T = 197°$ abs. lies about 10 per cent too low.[203]

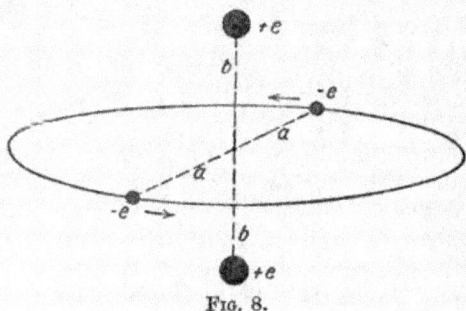

Fɪɢ. 8.

While all the above-mentioned investigations are based on the dumb-bell model, which can only be regarded as a provisional, schematic construction, *P. S. Epstein* [204] in 1916 carried out the corresponding calculations for another molecular model proposed by *N. Bohr*.[205] This model of the hydrogen molecule, to which we shall return later, is built up of two positive hydrogen atoms, each of which carry a single positive charge, and around the connecting line of which two electrons, diametrically opposite, rotate in a fixed circle at a fixed rate (see Fig. 8). Since the equilibrium in this purely electrical system is determined by the play of the *Coulomb* attractions and the centrifugal forces, and since the radius of the electron is determined by a quantum condition, this model possesses the advantage that all its dimensions are completely fixed, so that there is no longer any question of the arbitrariness of the moment of inertia. The "dumb-bell knobs" are represented here by the two positively-charged hydrogen atoms; the rotations of the molecule hitherto considered would therefore correspond to those motions in which the molecule rotates with a moment of inertia J about an axis at right angles to the line joining the atoms. But to this there must very plainly be added the rotation of the system about the axis of symmetry (i.e. the line joining the atoms), which results from the extremely rapid rotation of the electrons. The moment of inertia corresponding to this axis is, in consequence of the extremely small mass of the electrons, very small compared with J. The whole system obviously possesses, if we regard it approximately as rigid, the properties of a *symmetrical top*. Its motion is therefore, in consequence of its own rotation about the axis of symmetry, not a rotation, but instead the well-known motion, "regular precession."[206] *Epstein* treated the problem from this point of view but could not obtain agreement at low temperatures with the moment of inertia calculated from the model itself,[207] namely, $J = 2\cdot82 \times 10^{-41}$. Presumably, this failure depends on the fact that the model does not correspond with reality, and in fact we shall see later, that well-founded doubts have arisen as to the correctness of the *Bohr* model. We must therefore admit, unfortunately as one of a number of instances in the quantum theory, that the important problem of the rotational heat of hydrogen still awaits solution.

§ 2. The Bjerrum Infra-red Rotation-spectrum

N. Bjerrum[208] has applied the relation (79) in a very interesting manner to the infra-red absorption of polyatomic gaseous compounds. These gases (for example HCl, HBr, CO, H_2O in the form of steam, but on the other hand *not* the *elementary* gases H_2, O_2, N_2, Cl_2) show, according to the investigations of *S. P. Langley*,[209] *F. Paschen*,[210] *H. Rubens*,[211] *H. Rubens* and *E. Aschkinass*,[212] *H. Rubens* and *G. Hettner*,[213] *W. Burmeister*,[214] *Eva v. Bahr*,[215] extensive absorption bands in the short- and long-wave infra-red. While in the long-wave infra-red we account for the absorption by the rotating molecule, which, composed of positively and negatively charged atoms, act like electric double poles and hence in turning emit and absorb radiation, *Bjerrum* was the first to point out that the molecular rotation must also make itself noticeable in the short-wave infra-red. For if there exists in this region a linear vibration ν_0 of the ions in the molecule relatively to one another—and hence an absorption at this point—and if, in addition, the whole molecule rotates at the speed ν_r, then it is known that there will be produced as a result of the composition of the vibration with the rotation [216] two new vibrations (and, correspondingly, two new regions of absorption) having the periods $\nu_0 + \nu_r$ and $\nu_0 - \nu_r$, symmetrically disposed on both sides of the ionic vibration ν_0. On the whole, then, we have three points of absorption : $\nu_r, \nu_0 - \nu_r, \nu_0 + \nu_r$, to which we must add the non-rotational state ν_0 as a fourth. But if now, according to *Planck's* first theory, the molecule can only rotate with discrete speeds of rotation ν_n [see (79)], *we get symmetrically to the original position of absorption $\nu = \nu_0$ and, on both sides of it, a series of further discrete equidistant positions of absorption :*

$$\left.\begin{array}{l} \nu = \nu_0 + \nu_n = \nu_0 + n\,\dfrac{h}{4\pi^2 J} \\[2mm] \nu' = \nu_0 - \nu_n = \nu_0 - n\,\dfrac{h}{4\pi^2 J} \end{array}\right\} (n = 1, 2, 3 \ldots) \quad . \quad (83)$$

These discrete equidistant positions of absorption have actually been found by *Eva v. Bahr* in the case of water vapour and gaseous hydrochloric acid, and were measured later with still greater accuracy by *H. Rubens* and *G. Hettner* for water vapour. In an examination carried out on an extensive scale *E. S. Imes* [217] has once more thoroughly investigated the hydrogen halides (HCl, HBr, HF) and confirmed the law (83) for the position of the absorption lines. *It was thereby found that the middle line ν_0 was always missing.* From the standpoint of the theory here described this would mean that the non-rotational state does not exist, that is, *that the molecules always rotate* (zero-point rotation).

A. Eucken,[218] who discussed the results of *E. v. Bahr*, which were at that time the only ones known, deduced from the good agreement between observation and calculation that *Planck's* second theory is not valid, for the experiments seemed so obviously to prove that the molecule can actually only rotate with the discrete speeds ν_n. This conclusion, however, is not inevitable, as *Planck* [219] showed in a penetrating investigation. On the contrary, the observations may after all be explained, surprising as it may seem, on the basis of his second theory (continuous "classical" absorption; all speeds of rotation possible). This curious result is explained as follows: Let $w(E_r)dE_r$ be the probability that a molecule possesses exactly the rotational energy E_r; hence for N molecules $Nw(E_r)dE_r$ will be the number that will possess exactly the rotational energy E_r. These molecules rotate therefore—according to (78)—with the speed

$$\nu_r = \frac{1}{2\pi}\sqrt{\frac{2E_r}{J}}.$$

The quantity $w(E_r)$ is here, according to *Planck's* second theory, the *step-ladder curve* pictured in Fig. 7. *Planck's* calculation then leads to the following result: *the absorption of an external radiation of frequency ν^* is not—as one should expect—proportional to the number of molecules having a rotational speed $\nu_r = \nu^*$, that is, to the quantity $w(E_r)$ but to its differential coefficient* $\dfrac{dw(E_r)}{dE_r}$. This differential coefficient is, however, as Fig. 7 shows, everywhere equal to zero, excepting at the special quantum energy-values $E_r^{(n)}$, that is, at the rotational speeds ν_n. It follows from this, that here also, from the standpoint of *Planck's* second theory, *absorption takes place only at the special quantum rates of rotation ν_n.* It thus comes about that, at present at any rate, the infrared absorption spectra of the polyatomic gases, contrary to all expectation, do not decide one of the most fundamental questions of the whole quantum theory, whether, namely, *Planck's* first or second theory is correct. An important remark must be added here. The deductions of the relation which gives the position of the infra-red absorption bands is half in accordance with the classical and half in accordance with the quantum theory. For although the rotational speeds ν_n are determined by the quantum theory, the resolution of the oscillation ν_0 into the two components $\nu_0 \pm \nu_n$ are determined by the classical methods. How to attack this problem from a point of view entirely consistent with the quantum theory will be seen later in Chapter VIII.

§3. The Degeneration of Gases

The phenomena described above which were observed in the case of polyatomic gases (falling-off of the molecular heat, and infra-red absorption) justify fully the application of the quantum theory to motions of rotation. On the other hand, the attempts to go a step farther and to apply it to the *translational energy of gases* rest upon a much more insecure basis. If this step is taken, the hitherto exceptional position occupied by the monatomic gases, whose molecules contain only translational energy, becomes destroyed, for then they, too, must succumb to the quantum law. This problem has been attacked from various quarters [*O. Sackur*,[220] *H. Tetrode*,[221] *W. H. Keesom*,[222] *W. Lenz* and *A. Sommerfeld*,[223] *P. Scherrer*,[224] *M. Planck*.[225]] Thus, for example, *Tetrode, Keesom, Lenz* and *Sommerfeld* imagine the thermal motion of the gas split up into a spectrum of natural frequencies, and they then distribute the energy in quanta, that is, according to formula (9), over the individual natural frequencies, quite analogously to the manner of *Debye* and *Born-Kármán* in the case of solid bodies. *Scherrer* and *Planck*, on the other hand, apply the quantum hypothesis directly to the motion of the individual gas-atoms, basing their argument on the modern formulation of the quantum conditions for several degrees of freedom. How such a quantum resolution of the translatory motion is effected, is perhaps most easily seen by the following simple example (*Scherrer*): Let a gas-atom of mass m fly to and fro in a cube-shaped space of side a with the speed v parallel to one of the edges. It then executes a sort of oscillation with the period $\nu = \dfrac{v}{2a}$. If we set its kinetic energy, $E = \frac{1}{2}mv^2$, according to *Planck's* first theory $= n\dfrac{h\nu}{2}$ ($n = 0, 1, 2, 3 \ldots$) then it follows that

$$\tfrac{1}{2}mv^2 = n\frac{h}{2} \cdot \frac{v}{2a}$$

hence

$$v_n = \frac{nh}{2am} \quad \text{and} \quad E_n = \frac{n^2h^2}{8a^2m} \quad . \quad . \quad . \quad (84)$$

Hence the velocity of the atom and its translatory energy can acquire only discrete, quantum-determined values.

The calculations of the above-named investigators lead to two important main results, at least in qualitative agreement; in the first place, there results an alteration in the gas laws at very low temperatures. The necessity for this "degeneration" of the monatomic gases had already been recognised by *Nernst*, who deduced it on the basis of his new heat theorem.[226] For if the equation of state of ideal gases

$$pV = RT \begin{pmatrix} p = \text{pressure} \\ V = \text{volume of a gramme-atom} \\ R = \text{absolute gas-constant} \\ T = \text{temperature} \end{pmatrix} . \quad (85)$$

were exactly true for all temperatures down to the lowest, then the maximum work A, which could be gained from the isothermal expansion of the gas from the volume V_1 to the volume V_2, would have, as we know, for all temperatures the value

$$A = \int_{V_1}^{V_2} p\,dV = RT \int_{V_1}^{V_2} \frac{dV}{V} = RT \log_e \left(\frac{V_2}{V_1}\right).$$

For all temperatures down to absolute zero, $\frac{dA}{dT} = R \log\left(\frac{V_2}{V_1}\right)$

would *differ from zero*, in direct contradiction to the condition (38) of *Nernst's* Theorem. Hence it follows *that in the region of the lowest temperatures, the equation of state* (85) *must undergo modification.* In fact, experiments of *O. Sackur* [227] on hydrogen and helium appear to speak in favour of the existence of this "degeneration."

§ 4. The Chemical Constants of Monatomic Gases

The second main result given by the application of the quantum theory to monatomic gases, is an extremely interesting relation of the *Planck* constant h to the so-called "chemical constant" of the gas, a quantity which plays an important part in changes of the state of aggregation (vaporisation, sublimation) and in chemical states of equilibrium.   But it is here specially emphasised that the relationship just mentioned is not bound to the undeniably hypothetical resolution of the translatory energy into quanta. On the contrary, O. Stern [228] has succeeded in deducing it unobjectionably, *without applying the quantum theory to the gas.* The original method, which *Stern* adopts, may be shortly sketched here.   Consider the process of sublimation, i.e. the passage from the solid into the vapour state.   Let the vapour obey the gas laws, and let its density be negligible compared with that of the condensed solid.   Then classical thermodynamics gives for the pressure p of the saturated vapour as a function of the temperature the following equation :

$$\log p = -\frac{\lambda_0}{RT} + \tfrac{5}{2} \log T - \frac{1}{R}\int_0^T \frac{E_T^{(f)}}{T^2} dT + C \quad . \quad (86)$$

Here λ_0 is the heat of vaporisation (per gramme-atom) at absolute zero, $E_T^{(f)}$ is the energy of the condensed solid (per gramme-atom) at the temperature T; the constant C, which is the chemical constant of the vapourising substance, *remains undetermined, according to thermodynamics.*   On the other hand, the integral on the right-hand side of equation (86), which contains the energy of the solid material, may be completely calculated upon the basis of our assured knowledge of the energy-content of the solids. We only require to assume the solid to be a *Born-Kármán* crystal, and hence to use the quantum-theoretical value (41) for $E_T^{(f)}$.   If we now restrict ourselves to high temperatures, to a region, therefore, in which the classical theory is valid, (86) assumes the form

$$\log p = -\frac{\lambda_0 + \sum_1^{3N} \frac{h\nu^i}{2}}{RT} - \tfrac{1}{2} \log T + 3 \log\left(\frac{h\bar\nu}{k}\right) + C \quad (87)$$

6

(The $3N$ quantities ν_i here form the elastic spectrum of the solid body; $\bar{\nu}$ is their geometric mean.) The formulation of this equation constitutes the first step of *Stern's* deduction. It gives the result of thermodynamics, extended by the application of the quantum theory to the condensed substance. The second step is the formulation, in accordance with molecular theory, of a vapour-pressure formula for high temperatures, in the region therefore of classical statistics. Here, also, the *Born-Kármán* solid model is used for the condensed substance, and, on the basis of known laws of probability, the number of the atoms is calculated which are in statistical equilibrium in the vapour phase. In this way the density of the vapour, and hence, as a result of the gas laws, its pressure, are given. So *Stern* finds

$$\log_e p = -\frac{\lambda'_0}{RT} - \tfrac{1}{2} \log T + \log\left[\frac{(2\pi m)^{\frac{3}{2}}\bar{\nu}^3}{k^{\frac{5}{2}}}\right]. \quad (88)$$

Here m denotes the mass of an atom, and λ'_0 is the work which is necessary to bring N atoms (N is the *Avogadro* number) from complete rest to the gaseous state. An undetermined constant naturally does not appear in this formula deduced from pure molecular theory. For the molecular model is completely determinate, and hence gives the absolute value of the vapour pressure, not only its temperature coefficient, as in the case of thermodynamics. A comparison of (87) with (88) shows, firstly, that

$$\lambda'_0 = \lambda_0 + \sum_1^{3N} \frac{h\nu_i}{2} \qquad . \qquad . \qquad . \qquad (89)$$

and secondly, that the chemical constant C has the value

$$C = \log\left[\frac{(2\pi m)^{\frac{3}{2}}k^{\frac{5}{2}}}{h^3}\right] \qquad . \qquad . \qquad . \qquad (90)$$

Relation (89) may be interpreted by making the supposition that the solid body already possesses an energy amounting to $\sum_1^{3N} \frac{h\nu_i}{2}$ at the absolute zero, that is, a " zero-point energy," to which the latent heat of vaporisation λ_0 must be added, in order to set the atoms completely free from their union in the crystal. Equation (90) gives us the solution of the problem before us. *It gives the chemical constant of the monatomic gases as a function of the atomic mass and the universal constants h and k.* Nowhere, however, in the whole deduction—this should be emphasised once more—has the quantum hypothesis been applied to the gas itself.

In order to make formula (90) available for comparison with experiment [229] we may expediently introduce the molecular weight $M = mN$, and set $k = \dfrac{R}{N}$: then

$$C = C_0 + \tfrac{3}{2} \log M, \quad \text{where } C_0 = \log \left[\frac{(2\pi)^{\frac{3}{2}} R^{\frac{5}{2}}}{N^4 h^3} \right] = 10 \cdot 17$$

If we finally use the base 10 instead of the natural base e for our logarithms, and measure the vapour pressure not in absolute measure but in atmospheres, we get the chemical constant C' used by *Nernst*, which is related to *Stern's* value for C thus:

$$C' = \frac{C}{2 \cdot 3026} = 6 \cdot 0057$$

For it we finally get the simple expression

$$C' = C'_0 + \tfrac{3}{2} \log_{10} M, \quad \text{where } C'_0 = -1 \cdot 59 \quad . \quad (91)$$

This formula has been brilliantly verified by experiment. The hitherto most trustworthy measurements of vapour pressure and chemical states of equilibrium give in the case of hydrogen, argon, and mercury the values

$$-1 \cdot 69 \pm 0 \cdot 15, \quad -1 \cdot 65 \pm 0 \cdot 06, \quad -1 \cdot 62 \pm 0 \cdot 03$$

We are therefore justified in saying with *Stern* that the expression (90) for the chemical constant of the monatomic gases is theoretically and experimentally one of the best founded results of the Quantum Theory.

CHAPTER VI

The Quantum Theory of the Optical Series. The Development of the Quantum Theory for several Degrees of Freedom [230]

§1. The Thomson and the Rutherford Atomic Models

THE greatest advance since *M. v. Laue's* discovery of the method of Röntgen-spectroscopy for determining crystal structure was made in the realm of atomic theory in 1913, when the Danish physicist *Niels Bohr* placed the atomic models in the service of the quantum theory. *Bohr's* labours have in their turn reacted on the quantum theory and fertilised it, and thus a marvellous abundance of notable successes have been achieved in recent years through the interaction between the dynamics of the atom and the quantum hypothesis.

Among serviceable atomic models, the one proposed by *J. J. Thomson* long occupied a much favoured position; according to it, the electropositive part of an atom, which constitutes the most important part of its mass, is supposed to be a sphere of "atomic dimensions" (radius about 10^{-8} cms.) filled with a positive space charge in the interior of which the negative parts, the electrons, rest in a stable position of equilibrium. This model has the great advantage of explaining on purely electrical grounds the possibility of "quasi-elastically bound" electrons, i.e. such electrons as, being displaced from their position of rest, are drawn back into it by a *force which is proportional to the displacement*.[231] And it was just with the help of such electrons that, as is well known, *P. Drude*,[232] *W. Voigt*,[233] *M. Planck*,[234] and *H. A. Lorentz*[235] succeeded in building up large regions of theoretical optics, namely, the theory of dispersion and absorption, and the magneto-optical effects (magneto-rotation and *Zeeman* effect).

Moreover, the *Thomson* atomic model was able, by following the classical doctrine of the theory of electrons, to do what must be demanded of every serviceable atomic model, viz. to explain the emission, as a result of the oscillation of its electrons, of sharp, i.e. essentially monochromatic "spectral lines," the position of which, on account of the quasi-elastic restoring force,[236] was independent of the intensity of the excitation, that is, of the energy of the oscillations.

In three important points, on the other hand, the model failed completely. In the first place no success at all, unless with complicated and artificial hypotheses invented *ad hoc*, attended efforts to deduce from *Thomson's* model the formulæ for the optical series, for example, the simple formula for the *Balmer* series of hydrogen.[237] Secondly, the model could not account for the division of the spectral lines in an electric field as observed and closely studied by *J. Stark*[238] (*Stark* effect), in spite of the fact that it had been found most valuable, in the hands of *H. A. Lorentz*, for explaining and calculating the *Zeeman* effect.[239] Thirdly, it was not in a position to explain the large individual deflections, sometimes exceeding 90°, which, according to *H. Geiger* and *Marsden*,[240] α-particles undergo in passing through thin metallic foils. For on their way through the metallic foil, the α-particles, which are known to be doubly charged helium atoms, come into the neighbourhood of the metallic atoms and are more or less deflected from their straight paths by the electric fields of the atoms. If, now, the metallic atoms were *Thomson* atoms, the electric field of these atoms would attain its greatest value at the surface of the positive sphere, at a distance therefore of about 10^{-8} cms. from the centre of the atom. For from the surface outwards the field decreases, according to *Coulomb's* Law, with $\frac{1}{r^2}$, while it grows from the centre to the surface proportionately to r. Those α-particles, therefore, which pass close to the surface of the positive sphere, must undergo the greatest deflection. An easy approximate calculation shows, however, that the field at this distance from the centre is far from being strong enough to explain the great deflections which *Geiger* and *Marsden* have observed. This weighty reason led *E. Rutherford*[241] to set up, instead of the *Thomson* model, a new one, which was able to explain the large deflections of the α-rays. According to the *Rutherford* atomic picture, the electropositive part of the atom is compressed into an extremely small space [242] the so-called *nucleus*. Its charge E consists in general of z positive elementary charges e, so that $E = ze$. Here z is, according to a hypothesis of *van den Broek*,[243] the *atomic number* of the element, i.e. the number which gives the position of the element in the series of the periodic table. Thus, for example, $z = 1$ for hydrogen, 2 for helium, 3 for lithium, and so on. About this nucleus the electrons describe planetary paths, that is, circles or *Kepler* ellipses with the nucleus as focus, since the electrons are attracted by it in accordance with *Coulomb's* Law (inversely proportional to the square of the distance).

In the electrically neutral atom having the atomic number z, z electrons circle round the nucleus. For example, the neutral hydrogen atom consists of a singly charged nucleus ($E = e$) around which one electron revolves in a circular or elliptic path.

That this *Rutherford* model is actually able to explain the cause of large deflections of the α-particles is seen at once; for the field-strength of the nucleus, in contrast to *Thomson's* model, increases strongly up to the immediate neighbourhood of the nucleus, in accordance with *Coulomb's* Law; hence, if the positively charged α-particles come very close to the nucleus—that is, much nearer than 10^{-8} cms.—then they are exposed to the extremely powerful repulsion of the nucleus.

On closer examination, the *Rutherford* atomic model disappoints us seriously: for the revolutions per second, ν, of the electrons depend on the energy of the system.[244] If, therefore, we suppose, according to the classical electron theory, that an electron revolving at ν revolutions sends out an electromagnetic radiation of frequency ν, then, since the system loses energy by this radiation, ν must diminish correspondingly. But this means *that the atom is unable to emit a sharp, homogeneous spectral line.*

§ 2. Bohr's Model of the Atom

It thus appears that we are obliged to reject this model at the very outset. But the history of physics has decided otherwise. With deep-sighted intuition, *Niels Bohr* saw the possibilities of *Rutherford's* model and brought it under the quantum theory by making three bold hypotheses.[245] In the first place, he assumed that the electron (or electrons) cannot revolve around the nucleus in all paths possible according to the view of mechanics, but only in certain discrete orbits determined by the quantum theory. If we restrict ourselves, as *Bohr* did initially, to circular paths, then only those paths of an electron are allowable from the view of the quantum theory for which the moment of momentum (angular momentum) of the revolving electron is a whole multiple of $\frac{h}{2\pi}$, in exact agreement with the quantum condition (81) or

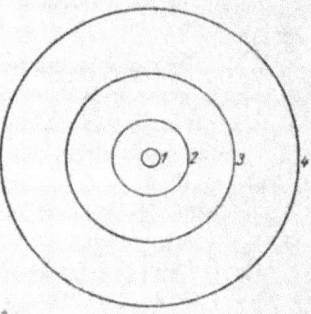

(82) for the rotating molecule. This gives, in the simplest case for the quantum paths of the electrons, a discrete family of concentric circles around the nucleus, with radii, which are related to one another as the squares of the whole numbers (1 : 4 : 9 : 16 : — —). See Fig. 9.

Secondly, these "allowable" orbits are stationary; they are in a certain sense

FIG. 9.

stable states of motion. This stability is gained by making the radical condition that the electron—in striking contrast with everything that the classical theory has taught us— shall not radiate when in the stationary paths. Since by this "decree" the loss of energy is abolished, the electron can continually revolve in such a "quantum path." That there are such "non-radiating" paths in the atom, is beyond doubt. Among other things, the constancy, in time, of the para- and ferro-magnetism of bodies, which is generated by revolving electrons, speaks in favour of this view. But how electrodynamics must be altered in order to guarantee the non-radiation of the quantum paths, and only of these, is a question which as yet remains unanswered. As we have now abolished the "classical" radiation of the atom, the actually observed emission must be accounted for by a new hypothesis.   Here, again in direct connexion with *Planck's* original quantum rule, *Bohr's* third condition takes effect: when the electron passes from one allowable quantum orbit, in which the energy is W_2, into another allowable quantum path of energy W_1, energy amounting to $W_2 - W_1$ is radiated in the form of an energy-quantum $h\nu$ of homogeneous, monochromatic radiation.   The frequency of the radiation emitted is determined by "*Bohr's* Frequency Condition:"

$$\nu = \frac{W_2 - W_1}{h} \qquad . \qquad . \qquad . \qquad . \quad (92)$$

We can follow *Einstein* [246] in imagining the passage from the state of higher energy to the state of lower energy as a sort of radio-active disintegration, the occurrence of which in time is determined by chance.   The details of this passage and the release of energy accompanying it are, however, entirely obscure up to the present.

§ 3. The Hydrogen Type of Series according to Bohr's Atomic Model

However bold and unorthodox *Bohr's* three hypotheses may have appeared, their success was surprising.   If we apply them to a "hydrogen type" of *Rutherford*-atom in which a *single* electron revolves around a positive nucleus with a z-fold charge, we get [247] for the frequencies of the spectral lines, which the electron emits in passing from the nth to the sth quantum path, the following values:

$$\nu = \frac{2\pi^2 e^4 m z^2}{h^3}\left(\frac{1}{s^2} - \frac{1}{n^2}\right) \quad \left\{\begin{array}{l} e,\, m \text{ charge and mass} \\ \text{of electrons} \\ s,\, n \text{ whole numbers} \end{array}\right. \quad (93)$$
$$= N z^2 \left(\frac{1}{s^2} - \frac{1}{n^2}\right)$$

If we here set $z = 1$ (hydrogen), $s = 2$, $n = 3, 4, 5 \ldots$ *we get in exactly the same form the empirical expression for the* Balmer *series of glowing hydrogen* [248]

$$\nu = N\left(\frac{1}{2^2} - \frac{1}{n^2}\right) \qquad (n = 3, 4, 5 \ldots) \qquad . \quad (94)$$

For the constant N which appears in the empirical formula, the so-called *Rydberg* number, *Bohr's* Theory therefore gives the expression

$$N = \frac{2\pi^2 e^4 m}{h^3} \qquad . \qquad . \qquad . \qquad . \qquad (95)$$

If we use here the well-known values
$e = 4\cdot774 \times 10^{-10}$ (*Millikan*) $h = 6\cdot55 \times 10^{-27}$ (*Planck*)

$$\frac{e}{mc} = 1\cdot77 \cdot 10^7$$

then it follows from (95) that

$$N = 3\cdot27 \cdot 10^{15}$$

while the empirical *Rydberg* number has the value $3\cdot29 \cdot 10^{15}$. This striking agreement and the resolution of the *Rydberg* number into universal constants is one of the main achievements of *Bohr's* Theory,[249] and forms a strong argument for its innate power. We may say that, according to *Bohr's* original theory, the individual lines of the *Balmer* series (H_a, H_β, H_γ, . . .) are emitted when the electron jumps from the 3rd, 4th, 5th . . . orbit into the 2nd.

With this statement, however, the achievements of formula (93) are not exhausted. For it includes, as we easily see, further spectral series of hydrogen. Namely, if we set $s = 1$, $n = 2, 3, 4$ . . . we get the ultra-violet series that was found and measured by *Lyman*.[250] If on the other hand we set $s = 3$, $n = 4, 5, 6$ . . . we get the infra-red *Bergmann* series, the first two lines of which were measured by *F. Paschen*.[251]

The element which follows hydrogen in the Periodic System is helium (atomic number $z = 2$). While, however, the constitution of the *neutral* helium atom with its two electrons is already more complicated—according to the latest investigations, the two electrons circle around the nucleus in two different orbits—the simply ionised helium atom, which has therefore a single positive charge, is entirely " of the hydrogen type;" for it consists of a doubly-charged positive nucleus around which an electron rotates. The sole difference, as compared with the hydrogen atom, thus consists in the doubling of the nuclear charge, $z = 2$. The series emitted from the positive helium atom may therefore, according to (93), be comprised in the formula

$$\nu = 4N\left(\frac{1}{s^2} - \frac{1}{n^2}\right) \qquad . \qquad . \qquad . \quad (96)$$

where N is again the *Rydberg* number as defined in (98). If we here set $s = 3,\; n = 4,\; 5,\; 6 \ldots$ we get the so-called "principal series of hydrogen" which was observed by *Fowler*[252] and very recently measured with great care by *F. Paschen*.[253] For $s = 4,\; n = 5,\; 6,\; 7 \ldots$ we get the so-called "second subsidiary series of hydrogen," which was observed by *Pickering*[254] and *Evans*.[255] Both series were, before the advent of *Bohr's* Theory, falsely ascribed to hydrogen.

A new and extremely noteworthy result of *Bohr's* Theory is revealed, if we allow for the movement of the nucleus in our calculations. For, in reality, the nucleus is not stationary, but nucleus and electron revolve about their common centre of gravity. By taking this fact into account we are led to a slightly altered expression for the *Rydberg* constant. In place of (95) we get the formula

$$N = \frac{2\pi^2 e^4 m}{h^3\left(1 + \dfrac{m}{M}\right)} \qquad . \qquad . \qquad . \quad (97)$$

in which M denotes the mass of the nucleus. It follows from this that for different elements, for instance, hydrogen and helium, the *Rydberg* constant differs somewhat and is smaller for hydrogen than for helium (since $M_H < M_{He}$). In general, the value of the *Rydberg* constant increases with increase of atomic weight tending towards a limiting value. All this is in perfect agreement with the results of many years of spectroscopic research.

In the same way as emission, *absorption* has a quantum-like character, according to *Bohr's* model. If light, say of the first *Balmer* line (H_a), falls upon a hydrogen atom, a quantum $h\nu$ of this external H_a radiation is used to "raise" the electron into the third quantum orbit. An amount of energy $h\nu_{H_a}$ is taken from the external radiation, that is, light from the line H_a is absorbed.

§ 4. The Structure of the Periodic System

Even in his earliest papers *Bohr* endeavoured to construct for the higher elements as well (Li, Be, B, C, etc.), in connexion with the Periodic System, suitable atomic models with several rings of electrons, each occupied by several electrons, in which, for example, the well-known octaves of the system are reproduced by a regular arrangement of the external electrons which recurs at every eighth element, while the number of the electrons revolving in the outermost ring is equal to the valency of the element in question.

W. Kossel [256] arrived at a similar structure of the atoms as a result of a profound investigation of the formation of molecules from atoms. Also, *L. Vegard*,[257] *A. Sommerfeld*,[258] and *R. Ladenburg* [259] have constructed analogous atomic models, particularly taking into account the well-known up-and-down curve of atomic volumes, and using them to explain other periodically varying properties (paramagnetism, ionic colour). These considerations, although they are tending indisputably along the right lines as far as the general principles are concerned, are not yet firmly established in detail.

§ 5. The Quantum Hypothesis for Several Degrees of Freedom

While the quantum hypothesis in its most primitive form demonstrated in this way its innate power by entering the field of atomic dynamics, it had, in doing so, gained little as far as its own development was concerned. But the fruits of *Bohr's* Theory ripened more rapidly than could have been divined. Already the year 1915 brought a decisive development: almost simultaneously, *Planck* and *Sommerfeld* independently found the solution of a problem that had long been a burning question, *namely, the extension of the quantum theory to several degrees of freedom*. *Sommerfeld* [260] retained a close connexion with *Bohr's* Theory in attacking the problem. The first main condition of this theory related to the choice of "allowable" stationary orbits among all those mechanically possible. According to this, as we saw, only those orbits were allowed for which the moment of momentum (*Impulsmoment*) p is a whole multiple of $\frac{h}{2\pi}$. This may also be

expressed according to (81) and (82) thus : among all mechanically possible paths, only those are allowable and stationary
for which the *phase-integral* fulfils the condition :

$$\int p\,dq = nh \qquad \qquad (98)$$

In this *quantum condition* we are to replace according to
(82) the general co-ordinate q by the angle of rotation (the
"azimuth") ϕ, the impulse p by the "impulse (or momentum)
corresponding to ϕ," namely, p_ϕ (the moment of momentum).
The integration is thereby to be extended over the whole range
of values of the variable q, that is, in the present case, from
0 to 2π.

In the case of the original *Bohr* Theory, which considers
only circular orbits, there naturally exists only a *single*
quantum condition, namely, that for the case $q = \phi$, since
the angle of rotation ϕ is the only variable of the path.
Matters are otherwise, when we reject the limitation to
circular orbits, and hence take *Kepler*-ellipses into account.
Then each point of the path is determined by two variables,
namely, by the distance r of the electron from the nucleus,
which is at the focus of the ellipse, and by the angle ϕ (the
"azimuth") which r makes with a fixed direction (say with the
straight line, which joins the nucleus to the perihelion). In
this case we are presented with a problem of two degrees of
freedom, with two generalised co-ordinates, r and ϕ (polar
co-ordinates). The simple extension of the quantum hypothesis by *Sommerfeld* now consists in setting up in this case
two quantum conditions of the form (98), one for the co-
ordinate ϕ, which agrees with the single quantum condition of
Bohr's Theory, and a new one for the co-ordinate r, so that
the selection of the stationary orbits is here determined by
the two following equations :

$$\int_0^{2\pi} p_\phi\,d\phi = nh . \qquad \qquad (99)$$

$$\int p_r\,dr = n'h . \qquad \qquad (100)$$

n and n' are here whole numbers, p_ϕ and p_r are the impulses
(momenta) corresponding to the co-ordinates ϕ and r.[261] The

integration in (100) is to be taken over the full range of values of r, that is, from the smallest value r_{min} (perihelion) to the greatest value r_{max} (aphelion) and back to the smallest r_{min}. (99) is called the azimuthal quantum condition, n being the azimuthal quantum number; (100) is the radial quantum condition, n' the radial quantum number.

In a corresponding manner the extension may be carried out for more than two degrees of freedom. If the system has f degrees of freedom, and if it is therefore characterised by the f generalised co-ordinates q_1, q_2, q_3 . . . and the corresponding impulses p_1, p_2, p_3 . . ., then the "allowable" movements of the system are limited by the f quantum conditions:

$$\int p_1 dq_1 = n_1 h; \int p_2 dq_2 = n_2 h, \ldots \int p_f dq_f = n_f h . \quad (101)$$

(n_1, n_2 . . . n_f are positive whole numbers).

In every one of the f phase-integrals the integration is to be performed over the full range of values of the co-ordinate in question.

A difficulty, which arose here from the outset, was the question as to *which* co-ordinates ought to be chosen for the application of the quantum rule (101), or whether the choice is immaterial. In general, we may characterise a system of several degrees of freedom by various types of co-ordinates; for instance, we may describe the *Kepler* movement of the electron either by polar co-ordinates r and ϕ, or by Cartesian co-ordinates x and y. This question is the more urgent, when one considers that the separate phase-integrals $\int p_i dq_i$ do not really become *constants* for every choice of co-ordinates, as is required by the quantum rule (101).[262] *P. S. Epstein*[263] and *K. Schwarzschild*[264] have solved, independently of one another, this problem of the "correct choice of co-ordinates" to a certain extent. Incidentally, an interesting and surprising relation of the quantum rules (101) to a long-known theorem of classical dynamics was revealed, which had been propounded by *Jacobi* and *Hamilton*, and had hitherto been successfully applied in celestial mechanics. Finally, quite lately, *A. Einstein*,[265] by modifying the expression (101), has put forward a quantum hypothesis which has the advantage of being independent of the choice of co-ordinates. But a closer discussion of these abstract investigations would lead us too far here.

The second formulation of the quantum hypothesis for several degrees of freedom is due, as already mentioned, to *M. Planck*.[266] It is, as it were, more cautious in its nature than the more radical attack of *Sommerfeld*. *Planck*, continuing directly from the division of the phase-plane of linear oscillators already discussed, starts from the so-called *Gibbs* phase-space to deal with more complicated systems. For a system of f degrees of freedom, which is characterised by the co-ordinates $q_1, q_2 \ldots q_f$ and the impulses $p_1, p_2 \ldots p_f$, the *Gibbs* phase-space is that $2f$ dimensional space, the points of which possess the $2f$ co-ordinates $q_1 \ldots p_f$. Each point of the phase-space (phase-points) represents, therefore, a definite momentary state of the system in question. *Planck* now gives this phase-space, in exact analogy to the phase-plane, a *cellular structure*, by bringing into prominence certain specially distinguished boundary surfaces. At the same time the size of the cells is proportional to h^f. The points of intersection of those boundary surfaces then represent the distinctive quantum states or phases of the system (that is, according to *Planck's* first theory the only possible, the "allowable" conditions). In contrast with *Sommerfeld's* Theory, in which the motion of a system of f degrees of freedom is always determined by f quantum conditions, in *Planck's*, under certain circumstances, the case may occur that fewer quantum conditions than degrees of freedom exist, so that several ("coherent") degrees of freedom are limited by a single quantum condition.

§ 6. Sommerfeld's Theory of Relativistic Fine-structure

That these theories had found the kernel of the matter was soon to be shown by applying them to *Bohr's* atomic model. According to them from among all the mechanically possible paths, which the electron can describe about the z-fold positively charged nucleus, the allowable, stationary paths must be determined by the two quantum conditions (99) and (100). This gives, in place of the discrete, quantised circles

of *Bohr*, *discretely quantised Kepler ellipses*, among which
also the *Bohr* circles are included, as special cases. And
further, the ellipses are quantum-determined, both with reference to their sizes (i.e. to their major axes), and to their
form (i.e. the relation of the axes to one another), so that here
every orbit, as compared with *Bohr*, is characterised by *two
quantum numbers* n and n'.[267] In place of formula (93) for
the hydrogen type of series, we get the general formula : [268]

$$\nu = Nz^2\left[\frac{1}{(s + s')^2} - \frac{1}{(n + n')^2}\right] \qquad . \qquad (102)$$

Here again N, the *Rydberg* constant, is given by (95), or
more exactly (the motion of the nucleus being taken into
account) by (97) ; s and s' are the two quantum numbers
(azimuthal and radial) of the final orbit of the electron ; n
and n' are the quantum numbers of its initial orbit. Since
also, as a result of this more complete view of *Sommerfeld*,
the number of allowable orbits is greatly increased, as compared with those arising from *Bohr's* Theory (owing to the
addition of the ellipses), the electrons have a great many
more possibilities in passing from one orbit to another, that is,
the chances of generating spectral lines are multiplied. But
we easily recognise the following fact : if we choose as the
final orbit of the electron any one of those orbits, for which
the sum of the quantum numbers s + s' has a definite value,
say s + s' = 2, and as initial orbit, any one of those paths,
for which n + n' has a definite value, say n + n' = 3, *then
all the different transitions of the electrons from any one of
these initial orbits to any one of these final orbits generate
always the same line* (in the case of the figures above chosen
it will be the first *Balmer* line) ; for according to (102) the
frequency of the line emitted depends only upon the sum
s + s', and the sum n + n', and on the other hand *not* on the
separate values of s, s', n, n'. It would thus appear as if
nothing is gained physically by *Sommerfeld's* elaboration of
the theory as compared with *Bohr's* original theory. However, as *Bohr* had already pointed out, the calculations are
incomplete in one important respect, which become of fundamental importance when consistently taken into account,
and which represents the main achievement of *Sommerfeld's*

theory of spectral lines. Namely, the velocities of the electrons, which appear in these problems, *cannot be considered negligibly small compared with the velocity of light*. In this case, however, we cannot, as we know, calculate by the methods of classical mechanics, which regards the mass of the electron as constant, but *must take our stand upon the theory of relativity*, and hence take into account the variations of the mass of the electron with its speed. *Sommerfeld* completed the calculation in this respect. The paths of the electron and the nucleus differ, in this refinement of the theory, from the ordinary *Kepler* ellipse in that the perihelion of the orbit advances in the course of time, and that the path loses its closed character. This has the effect that the energy of the electron in the stationary quantum-chosen orbits—which here also are determined by (99) and (100)—are no longer solely dependent on the sum of the quantum numbers as in the case of the non-relativistic *Kepler* motion, but that the quantum numbers n and n' also enter, separately, into the expression for the energy. Only as a first approximation, therefore, i.e. when the relativity correction is neglected, will the frequency ν of the spectral line emitted depend on the quantum sums $s + s'$ and $n + n'$ alone, as (102) shows. If we take into account the relativistic change of mass of the electron, on the other hand, ν will also depend on the individual values of s, s', n, n'.[269] *It follows, therefore, that the various possibilities, above considered, of the generation of a definite spectral line, that is, the passage of an electron from any one of the initial orbits $s + s' = $ constant to any one of the final orbits $n + n' = $ constant, no longer produce exactly the same line, but give rise to slightly different lines*, which, however, on account of the smallness of the relativity effect, lie very close together. This is *Sommerfeld's* explanation of the fine-structure of the spectral lines in the case of the hydrogen type of spectra. For example, according to *Sommerfeld*, the first line of the *Balmer* series (the red hydrogen line H_a) must consist of five components, which are arranged in two chief groups (of two and three each). The mean distance of these two groups from one another should amount, according to the theory,[270] to about 0·126Å; the best measurements of the hydrogen doublet gave the value 0·124Å (*Paschen, Meissner*).

If this agreement already speaks strongly in favour of *Sommer-feld's* Theory, the exact measurements, by *F. Paschen*,[271] of the' fine-structure of the lines of positive helium (*Fowler* series) have given a still more convincing proof of its correctness ; *almost without an exception, all the components required by the theory of the fine-structure appeared on the photographic plate, and thus proved strikingly the existence of the stationary paths of the electron and its relativistic change of mass.*

Two interesting consequences may yet be mentioned here ; they are directly connected with *Sommerfeld's* Theory and *Paschen's* observations. First of all they have rendered possible the use of the fine-structure measurements for a direct "spectroscopic" determination of the three funda-mental constants e, m_0 (mass of the electron at infinitely low speeds), and h.[272] Secondly, *K. Glitscher*[273] was able to show that we only find the spectroscopic observations, for example, the size of the hydrogen doublet, in agreement with the theory, when we use for the variation in the mass of the electron the formula given by the theory of relativity. On the other hand, *Abraham's* Theory of the rigid electron leads to formulæ which do not agree with experiment.

§ 7. Higher Elements

We thus see that *Rutherford's* atomic model as further developed by *Bohr* and *Sommerfield* far exceeded the ex-pectations which it could reasonably be expected to fulfil. At any rate, it has revealed to us the optical series of hydrogen and helium with undreamed-of precision as far as the finest details. But beyond these primary gains, it has undertaken a further series of successful attacks. Thus *Landé*[274] was successful in calculating the two series-systems of neutral helium (helium and parhelium) by taking, in contra-distinction to *Bohr*, a model of the neutral helium atom in which the two electrons circle around the double positive nucleus in two different orbits, either co-planar or else inclined at an angle to one another. In this case then, the external electron, the leaps of which generate the radiation, moves in a field in which the simple *Coulomb* Law no longer holds, on account of the disturbing influence of the inner electron. Examples of this type which differ from that of

7

hydrogen have been generally investigated by *Sommerfeld*, who has shown [275] that by giving up the *Coulomb* field we arrive, to a first and second approximation, at the *Rydberg* and *Ritz* forms of the series laws. A very promising beginning in setting up a quantum theory of the spectral lines was thus made.

§ 8. The Stark Effect and the Zeeman Effect in Bohr's Theory of the Atom

Under the circumstances the question forces itself upon us, whether the atomic model in its present state of development is able to account for the *Stark* effect, that is, the splitting up of the spectral lines as a result of the action of an external electric field on the electrons emitting the lines. For, as we may remember, the original *Thomson* model had completely failed just at this point. And how do matters stand as regards the *Zeeman* effect, the splitting up of spectral lines as a result of an external magnetic field? Could the new model explain these phenomena as well as the old? Both questions have fortunately been answered in the affirmative. As regards the *Stark* effect, *P. S. Epstein*,[276] in an important paper, succeeded in demonstrating the following: if we calculate the motion of the electron under the influence of the nucleus and the external field, according to the methods usual in celestial mechanics, and then choose from among all mechanically possible motions the allowable stationary orbits by applying the modern quantum rules for several degrees of freedom, and if, thirdly, we allow the electron to leap from one of these stationary paths into another (whereby we limit the infinite number of possible passages by a "principle of selection" presently to be discussed), then the *Bohr* frequency formula (92) gives with the most admirable accuracy and completeness, both as regards position and number, all the components of the resolved lines as observed by *Stark* in the cases of hydrogen and positive helium. This astonishing result must be regarded as a further strong support of the correctness of *Bohr's* model and its system of quanta. The theory of the explanation of the *Zeeman* effect has up to the present not been quite so successful. It is true that *Debye* [277] and

Sommerfield [278] have been able to derive the normal *Zeeman* effect (division of the original line into a triplet when the line of observation is perpendicular to the lines of force) by calculation from the model. The explanation, however, of two important phenomena in this field has not yet been accomplished : firstly, the anomalous *Zeeman* effect and its laws (*Runge-Preston* rule), and secondly, the fact, discovered by *Paschen* and *Back*,[279] that even in the case of lines with a complicated fine-structure, the normal triplet is formed as the magnetic field grows. Further investigation will, it may be hoped, unravel those difficulties.

§ 9. The Principles of Selection of Rubinowicz and Bohr

Inasmuch as the foregoing considerations deal only with the position of lines in the spectrum, i.e. with their *frequency*, we are still confronted with the problem of their form of vibration, i.e. their intensity and polarisation. Moreover, the important question had yet to be answered, whether all leaps of the electron from any one stationary path to any other are possible, or whether the number of allowable transitions must be limited by some "principle of selection." This also is, fundamentally, a question of intensity, for the position may be regarded as follows : the forbidden transitions correspond to zero intensity. The solution of this whole complex of problems has been greatly advanced quite recently. In the first place, *A. Rubinowicz*,[280] by applying the law of the conservation of the moment of momentum (*impuls-moment*) to the system atom + radiated wave, arrived at a principle of selection and a rule of polarisation of the following form : in atoms of the hydrogen type, which are removed from the influence of external fields of force, the azimuthal quantum number n of the electron [see formula (99)] can only alter by 0, $+1$, or -1, when emission takes place. In the first case, the light radiated is linearly polarised, in the two other cases circularly. The position of the plane of the orbit remains unchanged during the process of emission. In the case of atoms differing from the hydrogen type, and of more complicated structure, the position is less simple; if we set the *total* moment of momentum of all the masses forming part of the system (we know that this impulse remains constant during the motion), equal to a whole number, n^*, times $\frac{h}{2\pi}$, it is just the changes *in this number* n^* during the emission which must be limited by the principle of selection in the same manner, as, in the case above, the alterations in the azimuthal quantum number of the individual electron in its leaps were limited. Here also, zero change in the azimuthal quantum number gives linear polarisation, changes by ± 1, on the other hand, lead to circular polarisation. In place of the orbital plane we get the "invariable plane" (at right angles to the total moments of momentum or impulse-moments), the position of which in space remains unaltered. If, finally, the atom is exposed to an external field, say a homogeneous electric field (*Stark* effect) or a homogeneous magnetic field (*Zeeman* effect), then, as we know, only that component of the total turning impulse remains constant during the motion of the masses forming parts of the atom which is parallel to the external field. If we set these components of impulse $= n_1 \frac{h}{2\pi}$, then only the alteration of this number n_1 will be limited by the principle of selection (that is, the alterations must be $0, \pm 1$). The principle of selection is thus clearly weakened in its action by the external field, and can, if fields of irregular strength and direction act on the atom, become completely illusory, as, for example, in the case of electric discharges.

By means of entirely different considerations, *N. Bohr* [281] arrived at results which coincide, in essentials, with those of *Rubinowicz*, but exceed them greatly in range. *Bohr* started from the fact that in the limit for large quantum numbers, when the successive stationary states of the atom differ very little in the energy they involve, the frequency that the electron emits in its passage between neighbouring states becomes identical with the rate of revolution in the stationary orbit.[282] The electron therefore emits, according to *Bohr's* frequency condition, the same line that it sends out according to the classical theory of electrons. In other words, *for very high quantum numbers, the quantum theory passes over into the classical theory.* (*Bohr's* "Principle of Correspondence or Analogy.") Arguing from this principle, *Bohr* proceeds as follows : according to classical mechanics, the motion of the electron in *Bohr's* atom may be represented as the super-position of component harmonic vibrations of the frequency :

$$\nu_{kl} = \tau_1\omega_1 + \tau_2\omega_2 + \ldots + \tau_f\omega_f \quad . \qquad . \quad (103)$$

Here, $\tau_1 \ldots \tau_f$ are whole numbers which in general may have all values between $-\infty$ and $+\infty$; the $\omega_1 \ldots \omega_f$ are certain constants which depend on the character of the motion : f is the number of degrees of freedom. Let the amplitude of the partial vibration characterised by the numbers τ_1 to τ_f be $A_{\tau 1} \ldots A_{\tau f}$. Then, according to classical electrodynamics, ν_{kl} is the frequency of the radiated partial wave $(\tau_1 \ldots \tau_f)$ and $A_{\tau 1}^2 \ldots A_{\tau f}^2$ is a measure of its intensity. On the other hand, the following result is derived from the quantum theory (*Bohr's* frequency formula) for high quantum numbers : in the transition from an initial state characterised by the quantum numbers $m_1, m_2 \ldots m_f$ into a final state corresponding to the quantum numbers $n_1 \ldots n_f$, a line of frequency

$$\nu_{Qu} = (m_1 - n_1)\omega_1 + (m_2 - n_2)\omega_2 + \ldots + (m_f - n_f)\omega_f \ldots \quad (104)$$

is emitted. Here the quantities $\omega_1 \ldots \omega_f$ are the *same* constants as in (103). But, according to *Bohr's* Principle of Analogy, for high quantum numbers $\nu_{kl} = \nu_{Qu}$. Hence there follows from a comparison of (103) with (104)

$$\tau_1 = m_1 - n_1, \quad \tau_2 = m_2 - n_2 \ldots, \quad \ldots \tau_f = m_f - n_f \ldots \quad (105)$$

i.e. *the " classical " partial vibration* $(\tau_1 \ldots \tau_f)$ *corresponds to that quantum transition, in which the quantum numbers alter by exactly* $\tau_1 \ldots \tau_f$. *The polarisation and intensity of the wave emitted during this quantum transition may be calculated from the form of vibration and amplitude of the " corresponding classical " partial oscillation.* This principle which has been derived for high quantum numbers is extrapolated by *Bohr* with great boldness over the region of all quantum numbers. Thus the important " principle of correspondence " is obtained. If in the development of the electronic motion in terms of partial vibrations the term $(\bar{\tau}_1, \bar{\tau}_2 \ldots \bar{\tau}_f)$ is missing, then the corresponding transition

$$m_1 - n_1 = \bar{\tau}_1, \quad m_2 - n_2 = \bar{\tau}_2 \ldots, \quad m_f - n_f = \bar{\tau}_f$$

is not present. Hence there follows, for example, for atoms of the hydrogen type in a field free from force, the law *that the azimuthal quantum number can in all emissions only change by + 1 or − 1*, both of which lead to circularly polarised radiation. This law is somewhat more limited in form than that of *Rubinowicz*.

Both the principles of selection and the rules for the polarisation and the intensity have stood the test of comparison with experiment. *Rubinowicz* himself showed that his principle of selection and the rule of polarisation are in agreement with *Paschen's* measurements of the fine-structure of the helium lines, and further with the observations of the *Stark* effect and the normal *Zeeman* effect. *P. S. Epstein* [283] and *H. A. Kramers* [284] went still further, and were able to prove by profound investigations, based on *Bohr's* Theory, that the calculations of intensity along the lines sketched above were also in surprising agreement with observation. Finally, *Sommerfeld* and *Kossel* [285] in an interesting study have applied the *Rubinowicz* principle of selection to spectra differing from the hydrogen type as well, and have shown that it is able to explain why certain series appear more readily and are more favoured than others, as it were, and that, by the selection of the " possible " transitions, it sets a limit to the multiplicity of possible combinations in a manner which, so it appears, entirely agrees with experience.

§ 10. Collision of Electrons on the Basis of the Bohr Atom

While in this way, through the interpretation and unravelling of the universe and the almost bewildering abundance of spectroscopic observations, the conviction of the correctness of *Bohr's* atomic model deepened more and more, a series of observations of quite another kind became known and contributed considerably to the consolidation of *Bohr's* Theory. These were the investigations already mentioned earlier in connexion with the light-quantum hypothesis, which dealt with the collision of free electrons with gas molecules and atoms. These researches were conducted particularly by *J. Franck* and *G. Hertz* [286] and, in succession, by a considerable number of American investigators in a systematic manner. The manifold results of these interesting researches may be

sketched here schematically by a simple example. What
have we to expect when electrons collide with a *Bohr* Atom?
As a simple type of *Bohr* atom, let us choose a model in which
z electrons revolve around a z-fold positively charged nucleus
in stationary quantum paths. The nature and spatial arrangement of these paths, as well as the distribution of the electrons
among the individual paths will be left open, and we shall

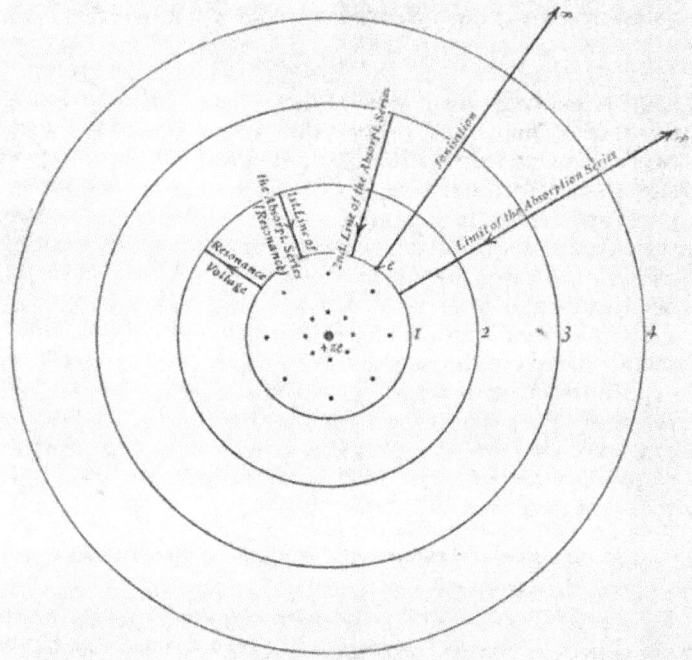

make only the simplifying assumption *that one electron*—the
so-called valency electron—*revolves alone in the outermost
orbit* (1) (see Fig. 10). Let this be the "normal," unexcited
state of the atom. The hydrogen atom ($z = 1$) is, as we know,
constituted in this way, and, of the neutral complicated atoms,
the atoms of the vapours of the alkali metals (Li, Na, K, Rb,
Cs) very probably also fall under this scheme. If by any
addition of energy the electron is "raised" from its normal

orbit (1) to a higher orbit (that is, one having more energy), say into the orbit (2), (3), (4) and so forth, and if it "falls" from these back into orbit (1), then the 1, 2, 3 . . . line of the so-called "*Absorption-series of the unexcited atom*" (principal series) is emitted. The frequencies of the lines emitted are regulated by *Bohr's* frequency condition (92), i.e. that the loss of energy $W_n - W_1$ incurred in passing from the nth to the first orbit is equal to a quantum $h\nu_{n,1}$ of the line emitted :

$$W_n - W_1 = h\nu_{n,1} . \quad . \quad . \quad . \quad (106)$$

The additional energy required to "raise" the electron to the higher energy level can be obtained in two ways : firstly by absorption of external radiation ; secondly (and that is the case we are dealing with here) by electronic impact. If external radiation of frequency $\nu_{n,1}$ falls upon the atom, a quantum $h\nu_{n,1}$ of this radiation is absorbed and is used to raise the electron from the energy level W_1 to the higher level W_n $= W_1 + h\nu_{n,1}$. In falling from this to the original level, the electron then emits the light corresponding to the line absorbed. The circumstance is further noteworthy, that the electron, when it is raised to the level (2), *has no other choice* than to return to the initial level, whereas from orbit (4) it can make one of three possible transitions—to (3), (2), and (1). If, therefore, the atom has absorbed light corresponding to the line $\nu_{2,1}$ from the external radiation, it will re-emit this line with its full complement of energy. The first line of the absorption series is, therefore, in contrast with all other lines, a so-called *resonance line*.

If the energy required to raise the electron is furnished by the impact of an outside electron, then—as *Franck* and *Hertz* were the first to prove—the intruding foreign electron will be reflected from the atom perfectly elastically (according to the mechanical laws of elastic impact), as long as its energy remains below a certain critical value E_R. If this energy value is reached, the impinging electron loses all its energy, and gives it up to the electron of the atom which has been struck ("inelastic impact"). What does this mean according to *Bohr's* view of the atom? Obviously E_R is nothing other than $W_2 - W_1$, that is, the energy which is necessary to raise the electron from its normal state in the atom to the orbit (2).

The result of this electronic impact, which adds energy of amount E_R to the atom must therefore be the emission of the resonance line. If this view represents the kernel of the matter, then the energy E_R must be connected with the frequency $\nu_{2,1}$ of the resonance line by the quantum relation

$$E_R = h\nu_{2,1} \quad . \qquad . \qquad . \qquad . \quad (107)$$

This relationship has been excellently verified by experiment. Thus *Tate* and *Foote*,[237] for example, find in the case of sodium, that the first inelastic electronic impact takes place when the impinging electron is accelerated by a potential of $V_R = 2\cdot2$ volts, the so-called resonance potential. The energy communicated by this potential to the impinging electron is

$$E_R = \frac{eV_R}{300} = \frac{4\cdot774 \cdot 10^{-10} \times 2\cdot12}{3 \times 10^2} = 3\cdot37 \cdot 10^{-12}$$

On the other hand, the resonance line that is under consideration here is the D-line, hence

$$h\nu_{2,1} = h\frac{c}{\lambda_{2,1}} = \frac{6\cdot545 \cdot 10^{-27} \cdot 3 \cdot 10^{10}}{5\cdot893 \cdot 10^{-5}} = 3\cdot33 \cdot 10^{-12}$$

We thus see that the relation (107) is fulfilled with great accuracy. The same holds for potassium ($V_R = 1\cdot55$ volts, $\therefore E_R = 2\cdot47 \cdot 10^{-12}$, $\lambda_{21} = 7\cdot685 \cdot 10^{-5}$ $\therefore h\nu_{21} = 2\cdot55 \cdot 10^{-12}$). In the case of the inert gases (helium, neon, etc.) and the vapours of mercury, zinc and cadmium, similar qualitative and quantitative relations—with some modifications—occur. The excitation, by electronic impact, of the mercury resonance line $\lambda = 2\cdot536 \cdot 10^{-5}$, that is $2\cdot536\overset{\circ}{A}$, discovered by *Franck* and *Hertz*, and already referred to, presents a characteristic example. The observed resonance potential is here $4\cdot9$ volts, while from the relation

$$V_R = \frac{300}{e}E_R = \frac{300}{e}h\nu_{2,1} = \frac{300hc}{e\lambda_{2,1}}$$

the value $V_R = 4\cdot86$ volts is deduced.

If the energy of the impinging electron is increased beyond E_R, then an "inelastic" impact, accompanied by complete loss of the energy, is to be expected every time as soon as E

has become equal to $W_n - W_1$ ($n = 3, 4, 5 \ldots$). By these various additions of energy the electron attached to the atom is raised successively to the 3rd, 4th, 5th . . . level of energy. If, finally, $E = E_\infty = W_\infty - W_1$, then the energy of the impinging electron is just sufficient to remove the electron attached to the atom to infinity, i.e. to *ionise* the atom. E_∞ is thus the ionisation energy, and the voltage corresponding to it, $V_\infty = \dfrac{300 E_\infty}{e}$, is called the *ionisation potential*. From the relation (106) we get immediately the important equation

$$E_\infty = \frac{e V_\infty}{300} = h \nu_{\infty 1} \qquad . \qquad . \qquad . \quad (108)$$

That is to say, the ionisation energy is equal to the quantum which corresponds to the last line of the absorption series, that is, to the "series limit." This quantum relation has also been excellently confirmed in all cases. For sodium, for example, *Tate* and *Foote* found: $V_\infty = 5\cdot 13$ volts, which gives an ionisation energy of the value $E_\infty = 8\cdot 17 \cdot 10^{-12}$. On the other hand, the limit of the principal series has the wavelength $\lambda_{\infty 1} = 2\cdot 413 \cdot 10^{-5}$, from which $h \nu_{\infty 1} = 8\cdot 14 \cdot 10^{-12}$, in striking agreement with the value of E_∞.

For mercury vapour, the limit in question of the principal series $\lambda_\infty = 1\cdot 188 \cdot 10^{-5}$. From this follows, according to (108), $V_\infty = 10\cdot 4$ volts while the measurements of various workers gave the value $10\cdot 2$ to $10\cdot 3$ volts (*Tate, Bergen, Davis* and *Goucher; Hughes* and *Dixon; Bishop* [288]). From all these examples, which could be considerably multiplied, the conclusion may be drawn with convincing clearness that the *Bohr* conceptions have laid bare the nature of the construction and the mode of action of the atom with unprecedented lucidity.

§ 11. Einstein's Deduction of Planck's Law of Radiation on the Basis of the Bohr Atom

Under these circumstances the suggestion naturally arises to refound the law of black-body radiation by taking as the elementary absorbing and emitting structure *Bohr's* model in place of the linear oscillator used by *Planck*. *Einstein* [289] has taken this step. In a highly important study he investigated the equilibrium of energy and momentum between black-body radiation and a generalised *Bohr* model, which, stripped of all special properties, has only to fulfil the quantum condition of being able to assume a discrete series of different states. For the interaction between the radiation and the atom—absorption (*Einstrahlung*) and emission (*Ausstrahlung*)—Einstein introduces the following simple hypotheses : the frequency of the emissions, i.e. the transitions, accompanied by loss of energy, of the atom from a condition (2) of higher energy, E_2, to a condition (1) of lower energy, E_1, shall follow the same statistical law as that which governs the disintegration of radioactive bodies, i.e. the number of transitions $2 \rightarrow 1$ in the time dt, or, as we may say, the number of atoms (2) that "disintegrate" in this time is proportional to $dt \cdot N_2$, where N_2 denotes the number of atoms momentarily in the state (2).

But, according to *Einstein*, a different law regulates the processes called into existence by the effect of external radiation. Under the influence of external radiation two things may happen : either an atom may pass from state (1) to state (2) by taking up energy, this is the "proper positive absorption." Or the case may also occur, that, as a result of the phase-relation between the field of the external radiation and the atom, the atom loses energy through the action of the impinging radiation, and hence passes from state (2) to state (1) ("negative absorption"). The rate at which both kinds of transition are repeated is then proportional to the intensity $\mathbf{K}_\nu$ of the external radiation : the number of transitions $1 \rightarrow 2$ associated with positive absorption in the time dt is therefore proportional to $N_1 dt \mathbf{K}_\nu$; the number of transitions $2 \rightarrow 1$ associated with negative absorption is proportional to $N_2 dt \mathbf{K}_\nu$. Here N_1 is the number of atoms momentarily in the state (1). N_1 and N_2 are determined by the laws of distribution known from the theory of gases and statistical mathematics and enlarged in conformity with the quantum theory. There follows from the energy equilibrium between in-coming and out-going radiation at the temperature T

$$\mathbf{K}_\nu = \frac{A}{e^{\frac{E_2 - E_1}{kT}} - 1} \qquad . \qquad . \qquad . \qquad (109)$$

where k is *Boltzmann's* constant, and A is a constant independent of the temperature. From *Wien's* Displacement Law (4) it follows, firstly that A is proportional to ν^3 and secondly that $E_2 - E_1$ is proportional to ν. If, therefore, we write

$$E_2 - E_1 = h\nu \qquad . \qquad . \qquad . \quad (110)$$

we recognise in this expression *Bohr's* frequency condition (92). In this way K_ν assumes the form of *Planck's* Law of Radiation, arising in a surprisingly simple and elegant manner from a minimum of hypotheses of a general character. *Einstein*, in pursuing and deepening these conceptions by writing down the expression for the equilibrium of the momenta in addition to the energies of the in-coming and out-going radiation, was led to the remarkable conclusion that the radiation of *Bohr* atoms cannot take place in spherical waves, as the classical theory of electrons requires, but that the process of emission must have a particular direction like the shot from a cannon. We cannot fail to recognise that this brings the conception that radiation has a quantum-like structure (light-quantum hypothesis) within realisable bounds.

CHAPTER VII

The Quantum Theory of Röntgen Spectra

§ 1. The Analysis of Röntgen Spectra

PARALLEL with the development of the science of optical spectra, a theory of Röntgen spectra has been developed of late years upon the same basis. This theory has already shed much light on the structure of atoms and thus forms a desirable extension of the theory of optical spectra. The investigations of *Ch. Barkla*, *W. H.* and *W. L. Bragg*, *Moseley* and *Darwin*, *Siegbahn* and *Friman*,[290] among others, have shown that by the impact of cathode rays upon the anti-cathode of a Röntgen tube two kinds of Röntgen rays arise: first, the so-called "impact radiation" (*Bremsstrahlung*) consisting of an extensive and continuous range of wave-lengths (similar to the continuous background of visible spectra); secondly, the "characteristic radiation," *a typical line-spectrum*, the structure of which depends so essentially on the material of the anti-cathode that a glance at this spectrum suffices us to deduce immediately and unmistakably the nature of the material of which the anti-cathode is composed. Thus alongside the optical spectrum analysis of *Bunsen* and *Kirchhoff* a Röntgen- or X-ray analysis presents itself. It has further been shown that the characteristic X-ray spectrum is a *purely atomic property, and, indeed, an additive one.* If we examine, for example, the X-ray spectrum, which is emitted by an anti-cathode of brass (copper + zinc), we find the lines of both copper and zinc unaltered and occupying the same positions as if only one metal were present in turn. No new lines appear. Accordingly we are led to suppose that the line-spectrum arises in the atoms of the anti-cathode, and is generated there by the impinging electrons of the cathode rays. The further important fact appeared that the lines of the characteristic spectrum *may be arranged in series*, just like those of the optical spectrum. Thus we have discovered up to the present a short-wave K-series, a long-wave L-series, and a still longer-wave M-series.

The most curious feature of these spectra is their connexion, by a definite law, with the atomic number of their element in the periodic system. If we plot the position of a certain line (say the first line K_a of the K-series) for the successive elements of the periodic system, a perfectly regular progressive shift is revealed : the line advances with increasing atomic number steadily towards the shorter waves. The regularity of this advance is such that we can recognise gaps or false positions of elements in the periodic system immediately by an excessive jump. Now, according to the hypothesis, already mentioned, of *Rutherford, v. d. Broek*, and *Bohr*, the atomic number of an element is nothing other than the number of its nuclear charge, that is, the number of elementary positive charges of its nucleus. If to this we add the phenomenon just discussed, according to which the steady advance of the nuclear charge in the series of the elements is reflected in the steady displacement of the X-ray lines, then we are forced to the view *that the origin of the X-ray spectra must be localised in the immediate neighbourhood of the nucleus, that is, in the inmost part of the atom.* For in this region the nucleus clearly has the greatest power and is least disturbed by external electrons, and hence it is here, too, that the growth of the nuclear charge will make itself most felt.

The connexion between the position of the X-ray lines and the atomic number z was first formulated by *G. Moseley*.[291] He found for the frequency of K_a (first line of the K-series) and L_a (first line of the L-series) the empirical relation

$$\left.\begin{aligned}\nu_{K_a} &= N(z-1)^2\left(\frac{1}{1^2} - \frac{1}{2^2}\right) \\ \nu_{L_a} &= N(z-7\cdot4)^2\left(\frac{1}{2^2} - \frac{1}{3^2}\right)\end{aligned}\right\} \qquad . \qquad . \quad (111)$$

where N is the *Rydberg* number.

The similarity of these relations, which are only approximately valid, with *Bohr's* formula (93) for the series of the hydrogen type is so striking, that it was an obvious step to seek to find the explanation of the Röntgen series by arguing on the basis of *Bohr's* model.

This problem was attacked chiefly by *W. Kossel*,[292] *A. Sommerfeld*,[293] *L. Vegard*,[294] *P. Debye*,[295] *J. Kroo*,[296] and *A. Smekal*.[297] And thus, in addition to the theory of the optical spectra which take their origin at the periphery of the atom, a theory of the Röntgen spectra has arisen which leads us

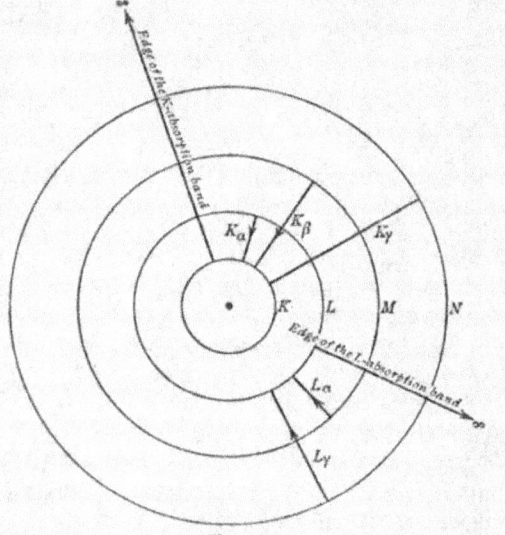

FIG. 11.

into the inmost regions of the atom. According to this theory we may picture to ourselves, in general terms, the emission of the Röntgen spectra as follows : we consider a neutral *Bohr* atom, consisting of a z-fold nucleus, around which z electrons revolve. These z electrons may be arranged in different rings. The innermost, single-quantum ring, the so-called K-ring, carries, let us say, p_1 electrons in its normal state ; let the second ring, the L-ring, be a two-quantum ring occupied by p_2 electrons, the third, three-quantum, the M-ring with p_3 electrons, and so on (Fig. 11). The question whether we can reach our goal with this conception of the ring by assuming the quantum numbers to increase as we go outwards, and whether we are to take the rings as co-planar or inclined to one another will be left open. The preparation for the emission of the K-series consists in this, that by the addition of energy—whether by absorption of external radiation or by electronic impact—an electron of the K-ring is removed to infinity, that is, the atom is, so to speak, ionised " inside," i.e. in the K-ring. If the energy of the atom before this inner ionisation $= W_0$, and after the ionisation $= W_K$, then the amount $W_K - W_0$ of energy must be provided. Hence every radiation, the energy quantum of which satisfies the condition $h\nu \geq W_K - W_0$, can on being absorbed effect the tearing of the electron out of the K-ring. If we allow the ν of the external radiation to grow slowly from small values, then, at the point $\nu_K = \dfrac{W_K - W_0}{h}$, a sudden increase of the absorption occurs, because from this point onwards the external radiant energy is used for the "ionisation of the K-ring." Thus an *absorption-band* extends from $\nu = \nu_K$ towards higher frequencies, the edge of the band lying at ν_K. This phenomenon of the " edge of the absorption-band " has already been interpreted above in the sense of the hypothesis of light-quanta. If the addition of energy is provided by the impact of a strange electron, coming from without, then its energy must be $E \geq W_K - W_0$, that is, $E \geq h\nu_K$, a relation, which we have already deduced earlier from the standpoint of the quantum hypothesis of light.

By ionisation of the K-ring the atom is now prepared for K-emission. If now an electron falls from the 2-quantum L-ring into the 1-quantum K-ring, filling up, so to speak, the gap produced there, then the first line of the K-series, K_a, will be emitted. If on the other hand the gap in the K-ring is filled by an electron of the 3-quantum M-ring, or the 4-quantum N-ring, K_β or K_γ result respectively. The position is quite analogous as regards the L- and M-series. If, by the addition of energy (absorption or electron-impact), an electron of the L-ring is battered off, that is if the L-ring is ionised, then the atom is prepared for the emission of the L-series. If, now, the gap in the 2-quantum L-ring is filled by an electron of the 3-quantum M-ring, the first line of the L-series,

L_{α}, results; if it is filled by an electron of the N-ring, the second line of the L-series, L_γ, results (the notation is not quite consistent but will serve the present purpose), and so forth.

The converse phenomenon to line emission, viz. line absorption, with which we are acquainted in visible spectra, appears at first sight to be missing here. That is, however, as $W.$ *Kossel* [298] recently showed, an error. It is true that the ejected electron of the K-ring, for example, cannot in general be caught upon the L-, M-, or N-ring, because all places on them are already occupied. An absorption of the lines K_α, K_β, K_γ, is therefore in this case impossible. But the electron of the K-ring can certainly come to rest on an *unoccupied* quantum orbit outside the occupied rings, that is, outside the surface of the atom. In this process a "line" is actually absorbed, namely, that line of which the $h\nu$ is equal to the energy-difference between the K-ring and the final orbit of the ejected electron. This refinement of our considerations shows, then, that the electron from the K-ring does not need to be raised immediately to infinity, but that line absorptions may occur before the edge of the band of absorption is reached.

§ 2. The Fine-structure of Röntgen Lines

It is particularly noteworthy that *Sommerfeld* succeeded also in the field of X-ray spectra in explaining the fine-structure of the lines by calling in the aid of the theory of relativity. Thus, for example, the 2-quantum L-orbit is "double"; it can occur as a circle ($n' = 0$, $n = 2$) or as an ellipse [299] ($n' = 1$, $n = 1$). Hence the line which is emitted by the electron of which the L-ring is the initial orbit, namely, K_α, is a doublet (K_α and $K_{\alpha'}$). In just the same way, those lines for which the L-orbit is the final orbit of the electron are doublets, namely, the line L_α (more exactly $L_{\alpha'}$) to which $L_{\beta'}$ is added to make a doublet; further, L_γ which forms a doublet with L_δ, and so forth. The distance between the components of the doublets (expressed in frequencies) comes out, according to *Sommerfeld's* Theory, as approximately proportional to the fourth power of the atomic number z. Hence here, in the X-ray region, where we are dealing for the most part with elements having fairly high atomic numbers, the

8

doublets appear macroscopically enlarged as compared with the microscopic hydrogen-doublet ($z = 1$). During the emission of X-rays the electron approaches very near to the highly-charged nucleus, and hence the relativistic effects of the resolution of the lines are much greater than in the case of the optical spectra, in which the electron is moving at the surface of the atom, where it is almost entirely screened from the action of the strong nucleus by the remaining electrons. With the help of the following relation deduced theoretically and adapted to experimental evidence,

$$\frac{L\text{-doublet}}{H\text{-doublet}} = (z - 3\cdot6)^4 \qquad . \qquad . \quad (112)$$

Sommerfeld was able to calculate the hydrogen-doublet from the observed L-doublets, and compare it with the results of experiment. The agreement is very satisfactory.

§ 3. The Distribution of Electrons among the Rings.   Objections to the Ring-arrangement of Electrons

The quantitative calculation of the simplest case, namely, the emission of K_a, led *Debye* to the conclusion that the K-ring in the normal state consists of three electrons. To this *Kroo*, by elaborating the calculation, adds the conclusion that the L-ring contains in its normal state nine electrons. With these two distribution numbers, $p_1 = 3$, $p_2 = 9$, the position of K_a could be represented as a function of the atomic number z for all elements. The emission of K_a takes place according to the following obvious scheme:

	K-ring	L-ring	
Normal state	3	9	
Initial state	2	9	} Ionisation of the K-ring.
Final state	3	8	} Emission of K_a.

The two distribution numbers (*Besetzungszahlen*) thus found for the two innermost rings excite our attention. For on the basis of the Periodic System with its periods of eight we ought to expect, according to *Kossel*, the numbers 2 and 8.

The strange occurrence of the numbers 3 and 9 becomes an objection, when we consider the case of sodium ($z = 11$). Here, according to *Kossel*, we should expect the numbers 2, 8, 1, since in all probability an electron (the valency electron) revolves alone, as in the case of all alkali metals, around the outside quantum orbit (M-ring). In any case it is impossible that the two innermost rings together should, in the *normal state*, contain 12 ($= 3 + 9$) electrons. If we attempt to go a step further still on the basis of *Kroo's* numbers 3 and 9, and to set up a formula which represents for all z's the position of L_a in conformity with observation, and thereby to determine the number of electrons p_3 on the M-ring, we find, as *A. Smekal*[300] showed, that this mode of representation is impossible with any combination 3, 9, p_3. Nor do we fare better if we incline the various rings to one another, and take their interaction into account. The suspicion is forced upon us, that perhaps the whole conception of the arrangement into plane rings does not correspond with fact, but that, rather, the electrons in the atom form *spatially* symmetrical figures. This suspicion is very much strengthened by a series of profound investigations carried out by *M. Born* and *A. Landé*.[301]

Following on *M. Born's* investigations of the dynamics of the crystal-lattice, which we discussed in detail earlier in connection with the atomic heat of solids, the two investigators asked themselves the question, whether it is possible to build up the cubic crystal-lattice of the alkaline halides (NaCl, NaBr, NaI; KCl, KBr, KI, etc.) from ions of *Bohr* atoms, by taking into account only the mutual electrostatic forces; and whether this method, if possible, would enable them to prophesy the crystal properties (lattice-constant, compressibility) from the atomic models of the two constituent ions. The answer to this question has been, on the whole, in the affirmative. But when the calculation of the compressibility of these crystals was carried out, the remarkable result manifested itself that crystals are found to be too soft, that is, insufficiently rigid, if the conception of the ring-arrangement of electrons in the atom is maintained. On the other hand, we get good agreement with the observations if, following *Born*, we introduce the hypothesis that the electrons are arranged *spatially*. A complex of eight electrons, as occurs in sodium, potassium, etc., does not therefore occupy a plane 8-ring; the eight electrons describe paths of *cubical symmetry*. Into the still obscure region of these " spatial " electron paths, *A. Landé* [302] has made some successful incursions.

From all that has been said it would appear to be certain that *in dealing with Röntgen spectra, too, we can no longer be content with the arrangement of the electron rings in planes*, and that the whole quantitative theory of the Röntgen series, including *Sommerfeld's* fine-structure of the *K*- and the *L*-doublets, must be built up on a fresh foundation.

CHAPTER VIII

Phenomena of Molecular Models

§ 1. Dispersion and Magneto-rotation of the H_2 Molecule

WHILE the X-ray spectra and the spectra of the optical series arise from the *atoms* of the elements (and hence their theory links up with the *atomic models*), there is a series of phenomena which, in the case of polyatomic substances, are peculiar to the molecules, and the theory of which, therefore, is founded on the molecular models. Chief among these are the normal dispersion, the rotation of the plane of polarisation in the magnetic field (magneto-rotation), and, further, the great and complicated subject of *band-spectra*. Up till a few years ago, dispersion and magneto-rotation had been exclusively treated from the standpoint of the *Thomson* model, that is, with the help of quasi-elastically bound electrons, and this explanation had served in turn as a powerful support for this model. Nevertheless, discrepancies in these theories had long been known. For example, measurements calculated upon the basis of the dispersion theories of *Drude*, *Voigt*, or *Planck* led to values for the ratio of the charge to the mass of the electron $\left(\dfrac{e}{mc}\right)$ which, in comparison with the direct measurements of this quantity (based upon the deflection of the cathode- or β-rays in the electric and magnetic fields) which were *much too small*. When, however, the *Thomson* model became displaced by the *Rutherford-Bohr* model, and the successes of the *Bohr* atomic model increased at an undreamed-of rate, the question arose whether an unobjectionable theory of dispersion and magneto-rotation could not be founded upon these new views. The difficult position, into which we are brought by this problem,

117

arises from the fact that we do not actually know a single instance of the exact manner in which a polyatomic *Bohr* molecule is built up from its nuclei and electrons. The exact knowledge of this structure, and the motion of all the electrons is absolutely necessary, if we desire to know how the molecule reacts upon external waves (dispersion). It is true that *W. Kossel*[303] has, in a detailed study already referred to above, pointed out the general guiding lines along which, from the chemical point of view, the building-up of the atom from molecules must be carried out, but the details of this construction remain open. Only in a few of the simplest cases have detailed molecular pictures been constructed and closely tested. Thus *Bohr*, as we remarked in discussing the atomic heat of gases, has already proposed a model of the diatomic hydrogen molecule. It has the following construction (see Fig. 8): two singly-positive nuclei (that is, each consisting of only a single positive charge) are separated by the distance $2b$. In the vertical plane which bisects the line joining the nuclei, two electrons rotate, diametrically opposite one another, on a circle of diameter $2a$. The equilibrium of the Coulomb and the centrifugal forces requires that $a = b\sqrt{3}$. By means of this relation, and by the quantum condition that each electron must have the moment of momentum $\frac{h}{2\pi}$, the model is completely determined in all its dimensions and speeds. It was this model which was the first to be proposed: it was examined by *P. Debye*[304] with reference to its dispersion. On account of its symmetrical structure the molecule possesses no electrical moment in its normal state. If, on the other hand, it is struck by an external light wave, the motion of its electrons is periodically disturbed; they depart from the normal quantum path, fall into forced vibration, and thus generate an electric moment which changes periodically in step with the external wave. Thus the original motion of the primary wave is changed, and dispersion results. We may conceive this as follows: Let c be the velocity of the primary wave *in vacuo*. The oscillations of the electrons generate a secondary wave which spreads out from the molecules. All these secondary waves combine with the primary wave to a form new wave which moves with the altered velocity q, the value of which depends on the frequency of the primary wave. But just this is the phenomenon of dispersion. The electronic vibrations which occur here are not oscillations about positions of equilibrium, as in the case of the quasi-elastic model, but oscillations about *stationary paths*. Moreover, here, the force holding the electrons, as opposed to the usual classical theories of dispersion, is *anisotropic* (that is, the electron is held by different forces in different directions); *above all, by means of this anisotropy, it was possible to explain away the disagreement in the value of* $\frac{e}{mc}$, *which had previously been found to be too small;* and *Debye* succeeded, on the basis of the normal value of $\frac{e}{mc}$, in deducing from the theory the observed dispersion curve of hydrogen, that is, the curve which shows how its coefficient of refraction depends on the wave-length. It should be noted that in the formula for the coefficient of refraction, no single constant is arbitrary, but that the dispersion formula is made up entirely of universal constants.

Using the same method (calculus of disturbances), *P. Scherrer* [305] has calculated the rotation of the plane of polarisation which linearly polarised light undergoes in its passage through hydrogen under the influence of a magnetic field. His efforts were equally successful.

§ 2. Objections to Bohr's Model of the Hydrogen Molecule

In spite of the successes which the *Bohr* model of the hydrogen molecule has won, a list of weighty objections to it has accumulated in the course of time. That the contribution which the rotation (more accurately, the regular precession) of this molecule makes to the molecular heat at low temperatures, does not correspond with the observations of *Eucken*, has been shown by *P. S. Epstein*, as we have already mentioned. Also at high temperatures, when the oscillations of the two nuclei relatively to one another contribute to the molecular heat, no agreement between theory and observation has been found in the case of the *Bohr* model, as *G. Laski* [306] recently showed.

Further, the model must possess, in consequence of the revolving electrons, an almost fixed magnetic moment parallel to the axis of the nucleus, that is to say, it must be equivalent to a molecular elementary magnet, which endeavours to set itself, in an external magnetic field, parallel to the lines of force. Hydrogen ought, therefore, to be paramagnetic, whereas it is diamagnetic.

Another very important objection, to which *Nernst* in particular drew attention, is the following : if we calculate the work which is necessary to separate the molecule into its two atoms, the so-called *heat of dissociation*, we get [307] the value, 61,000 calories. On the other hand, *Langmuir* [308] found 84,000 cals., *Isnardi* [309] 95,000 cals., *J. Franck*, *P. Knipping* and *Thea Krüger* [310] 81,000 (± 5700) cals. In any case, the calculated heat of dissociation comes out 25 per cent. too small.[310a]

Finally, *W. Lenz* [311] has recently increased the objections to the hydrogen model by an important one based on a theory of band-spectra, which we shall discuss below. He proved that the band-lines of hydrogen and nitrogen can exhibit the observed *Zeeman* effect, *only if these molecules possess no moment of momentum around the nuclear axis.* The fact that the two electrons in *Bohr's* molecular model revolve in the same sense, however, endows it with just such a moment of momentum. *On the whole, the Bohr model does not seem to correspond to reality ;* the arrangement of the two nuclei and electrons must plainly be quite different. No satisfactory model, however, has yet been found.

§ 3. Models of Higher Molecules

Matters are no better in the case of models of the more complicated molecules. It is true that *Sommerfeld* [312] and *F. Pauer* [313] have also worked out the theories of dispersion and magneto-rotation in the case of the more general *Bohr* models (N_2 and O_2) which are constructed on the lines of the hydrogen model. According to *Sommerfeld*, four electrons revolve about the line joining the two nuclei in the case of oxygen, each of which acts with an effective charge $+ 2e$; in the case of nitrogen, a ring of six electrons rotates about the nuclear axis, while the nuclei carry triple effective charges.

Sommerfeld was able to obtain agreement with observation only by setting up for each electron of a valency ring of $2s$-electrons the unaccountably strange quantum condition : moment of momentum $= \frac{h}{2\pi} \sqrt{s}$, undoubtedly a most unsatisfactory result. *Gerda Laski* [314] obtained better results with somewhat different models, which she chose in such a way that the specific heat of the two gases at high temperatures agreed with the observations of *Pier*.[315] According to her ideas, the nitrogen molecule must consist of two seven-fold positive nuclei, each of which is closely surrounded by a 1-quantum ring of two (or three) electrons. The "valency ring" in the central vertical plane is 2-quantum and contains ten (or eight) electrons. Analogously, the oxygen molecule consists of two eight-fold positive nuclei, each encircled by a 1-quantum ring of two (or three) electrons, whereas the 2-quantum valency ring contains twelve (or ten) electrons. The same objections apply to some extent to these models of *Sommerfeld* and *Laski* as to the hydrogen model. For example, they give no account of why oxygen should be paramagnetic, and nitrogen, on the other hand, diamagnetic. Moreover, the above-mentioned objection of *Lenz* applies in full force to these models; for they all possess moments of momentum around the nuclear axis. In conclusion, we feel bound to admit that the exact constitution of even the simplest models is at present unknown to us.

§ 4. The Quantum Theory of Band-spectra

To conclude this chapter, we shall turn our attention to the *band-spectra*, and collect together shortly what the quantum theory has been able to assert about them up to the present time. That they belong to molecules and compounds may nowadays be regarded as certain. The first attempt to construct a logical quantum theory of band-spectra was undertaken by *K. Schwarzschild* [316] who clearly recognised the importance of the rotation of the molecule in the production of these spectra. His conceptions may be defined as follows : a system of electrons revolves at a definite quantum distance around a molecule which itself rotates according to quantum conditions, the assumption being made for the sake of simplicity that the motion of the electrons is not influenced by the motion of the molecule. If E_0 is the quantum energy of the electrons, E_r the quantised rotational energy of the molecule, then $E_0 + E_r = E$ is the total energy of the system. If the three chief moments of inertia of the molecule J are equal to one another, then it follows, just as in (80), that

$$E_r = \frac{n^2 h^2}{8\pi^2 J}$$

where n denotes the rotational quantum number. Therefore

$$E = E_0 + \frac{n^2 h^2}{8\pi^2 J} \qquad . \qquad . \qquad . \quad (113)$$

If, now, the system passes from one quantum state having the electronic energy E_0 and the rotational quantum number n into another quantum state having the electronic energy E'_0 and the rotational quantum number n', then it follows from *Bohr's* frequency formula (92) that the frequency of the line radiated is given by

$$\nu = \frac{E_0 - E'_0}{h} + \frac{(n^2 - n'^2)h}{8\pi^2 J} \qquad . \qquad . \quad (114)$$

If we keep all the quantum numbers which occur here, excepting n, constant, and allow n to vary, then we get a series of lines progressing towards the violet and having the frequencies

$$\nu = a + bn^2 \qquad (a \text{ and } b \text{ are constants}) \qquad . \quad (115)$$

This is a formula which had already been given empirically by *Deslandres*,[317] and which is approximately true for the lines of many bands.

Following *Schwarzschild*, *T. Heurlinger* [318] and *W. Lenz*,[319] in particular, have further developed and refined the quantum theory of band-spectra. For example, *Lenz* has pictured the molecule as a symmetrical top having two moments of inertia and a rotational rigidity (moment of momentum) around the axis of the figure, and hence deals from the outset with a regular precession of the molecule in place of a rotation. Using *Bohr's* frequency formula, and applying the principles of selection, he obtained the following general formula for the lines of a band :

$$\nu = a + bn + cn^2 \qquad (a, b, c \text{ are constants}) \qquad . \quad (116)$$

which is obeyed, according to *Heurlinger*, in the case of the so-called "cyanogen" lines of nitrogen, for example. In addition to the lines given by (116), *Lenz's* Theory requires the occurrence of the series given by the formula

$$v = \left(a + \frac{b}{2}\right) + cn^2 \quad . \quad . \quad . \quad (117)$$

for the case *that the molecule really possesses a finite moment of momentum about its axis of figure.* A series which follows this law does *not*, however, exist in the cyanogen bands, according to *Heurlinger*. *Lenz* deduces from this the conclusion already mentioned, that the nitrogen model does not possess a rotational rigidity about its axis. By calculating the *Zeeman* effects of the band lines, and comparing them with observation, *Lenz* was able to confirm this, and to extend it to the hydrogen molecule.

The infra-red *Bjerrum* absorption bands of the diatomic and polyatomic gas compounds, which we had discussed at length in Chapter V, belong to the general type of band-spectra. If we are to deduce them from a theory consistently founded on quanta—and not, as we did earlier, half according to the quantum, half according to the classical theory—we must follow closely the course pursued above, with the difference that, in place of the energy of the electronic system there will appear the *energy of the atoms*,[320] with which the rotational energy of the molecule is combined, as a first approximation, additively. The logical carrying out of this calculation (in which *Bohr's* frequency formula and the principle of correspondence are applied), which was undertaken by *Heurlinger*[321] and the author,[322] gives for the structure of the "fluted" absorption bands an arrangement of lines which at first sight does not appear to agree with the beautiful and exact measurements of *Imes*.[323] The theory gives for the position of the absorption lines a formula

$$v = v_0 \pm (n + \tfrac{1}{2})\frac{h}{4\pi^2 J}, \quad (n = 1, 2, 3 \ldots) \quad (118)$$

and therefore requires that *all* neighbouring lines be equidistant, including the *two in the middle* ($n = 0$). On the other hand, *Imes'* observations show with indubitable clearness that the interval between the two middle lines is *twice* as great as the interval between all neighbouring lines. This apparent contradiction is explained, as *A. Kratzer* [324] recently showed, in a surprising fashion, if we take into account the *intensity* of the absorption lines according to *Bohr's* Principle of Analogy. For it then appears that the first absorption line to the right of the middle ν_0-line, namely, the line

$$\nu = \nu_0 + \frac{h}{8\pi^2 J}$$

(which is derived from formula (118) by setting $n = 0$ and using the positive sign for the second term) is of *vanishingly small intensity*. This line is generated when the molecule passes over from an initial rotationless and vibrationless state into the final state in which the two ions oscillate relatively to one another with one quantum, and in which, at the same time, the molecule rotates as a whole with one quantum. The rotationless and vibrationless state has, however, a vanishingly small probability; the number of transitions from this initial state per second, and therefore the intensity of the corresponding absorption line, is hence vanishingly small. By the disappearance of the first line to the right of the middle position ν_0, the structure of the lines as observed by *Imes* is actually reproduced, as one may easily recognise; in the formula, the "middle" of the line structure is displaced from the point ν_0 to the right by the amount $\frac{h}{8\pi^2 J}$. The absorption lines group themselves equidistantly and symmetrically on both sides of the missing "middle," $\nu = \nu_0 + \frac{h}{8\pi^2 J}$. This state of affairs may be expressed by writing, in formal agreement with (83),

$$\left.\begin{aligned} \nu &= \nu'_0 \pm \frac{nh}{4\pi^2 J} \quad (n = 1, 2, 3 \ldots) \\[2mm] \text{where} \\[1mm] \nu'_0 &= \nu_0 + \frac{h}{8\pi^2 J} \end{aligned}\right\} \qquad . \quad (119)$$

From the constant interval between neighbouring lines, namely

$$\Delta \nu = \frac{h}{4\pi^2 J} \quad . \qquad . \qquad . \quad (120)$$

the moment of inertia of the rotating molecule can be calculated with great accuracy.[325]

CHAPTER IX

The Future

IN the preceding pages the author has attempted to give in broad outline the most important features of the doctrine of quanta, its origin, its development, and its ramifications. If we now survey the whole structure, as it stands before us, from its foundations to the highest story, we cannot avoid a feeling of admiration; admiration for the few who clear-sightedly recognised the necessity for the new doctrine and fought against tradition, thus laying the foundations for the astonishing successes which have sprung from the quantum theory in so short a time.

None the less, no one who studies the quantum theory will be spared bitter disappointment. For we must admit that, in spite of a comprehensive formulation of quantum rules, we have not come one step nearer to understanding the heart of the matter. That there are discrete mechanical and electrical systems, characterised by quantum conditions and marked out from the infinite continuity of " classically " possible states, appears certain. But where does the deeper cause lie, which brings about this discontinuity in nature? Will a knowledge of the nature of electricity and of the constitution of the electromagnetic field serve to read the riddle? And even if we do not set ourselves so distant a goal, there remains an abundance of unanswered questions. The decision has not yet been made, as to whether, as *Planck's* first theory requires, only quantum-allowed states exist (or are stable), or whether, according to *Planck's* second formulation, the intermediate states are also possible. We are still completely in the dark about the details of the absorption and emission process, and do not in the least understand

125

why the energy quanta ejected explosively as radiation should form themselves into the trains of waves which we observe far away from the atom. Is radiation really propagated in the manner claimed by the classical theory, or has it also a quantum character?

Over all these problems there hovers at the present time a mysterious obscurity. In spite of the enormous empirical and theoretical material which lies before us, the flame of thought which shall illumine the obscurity is still wanting. Let us hope that the day is not far distant when the mighty labours of our generation will be brought to a successful conclusion.

Mathematical Notes and References

1 *O. Lummer* and *E. Pringsheim*, Wiedem. Ann. **63**, 395 (1897) ; Verhandl. d. deutsch. physikal. Ges. 1899, pp. 23, 215 ; *ibid.*, 1900, p. 163. Cf. also *O. Lummer* and *E. Jahnke*, Drudes Ann. **3**, 283 (1900), and *O. Lummer, E. Jahnke* and *E. Pringsheim*, Drudes Ann. **4**, 225 (1901).

2 Cf. *M. Planck*, Vorlesungen über die Theorie der Wärmestrahlung (Leipzig 1906), § 10.

3 Frequency $(\nu) = \dfrac{\text{velocity of light in vacuo } (c)}{\text{wave-length in vacuo } (\lambda)}$.

4 Cf., for example, *M. Planck*, Vorlesungen über Wärmestrahlung (1906), § 17.

5 *G. Kirchhoff*, Gesammelte Abhandlungen (J. A. Barth, Leipzig 1882), pp. 573 *et seq.* ; Berliner Akademieberichte, 1859, p. 216 ; Poggend. Ann. **109**, 275 (1860).

6 *O. Lummer* and *W. Wien*, Wiedem. Ann. **56**, 451 (1895). Cf. also *O. Lummer* and *F. Kurlbaum*, Verhandl. d. deutsch. physikal. Ges. **17**, 106 (1898).

7 Cf. Note 5.

8 *L. Boltzmann*, Wiedem. Ann. **22**, 291 (1884).

9 *J. Stefan*, Wiener Ber. **79**, 391 (1879).

10 The *Stefan-Boltzmann* Law is deduced as follows : Let the energy of black-body radiation at the temperature T, which is enclosed in a space of volume V having a movable piston, be $U = Vu$, where u is the "spatial" density of the radiant energy. The pressure, equal in all directions, which the radiation exerts upon the piston and walls is, according to electrodynamics, $p = \frac{1}{3}u$. If we supply to this system at the temperature T (that is, isothermally) an amount of heat $d'Q$, then its energy increases by dU, and the radiation does work pdV in pushing back the piston. Therefore, according to the first law of thermodynamics, and owing to the two relations above :

$$d'Q = dU + pdV = udV + Vdu + \frac{u}{3}dV = \tfrac{4}{3}\,udV + Vdu.$$

According to the second law of thermodynamics, $\dfrac{d'Q}{T}$ must be a complete differential. Hence the following relation holds :

$$\frac{\partial}{\partial u}\left(\frac{\frac{4}{3}u}{T}\right) = \frac{\partial}{\partial V}\left(\frac{V}{T}\right) \quad \text{i.e.} \quad \frac{4}{3}\frac{d}{dT}\left(\frac{u}{T}\right)\frac{dT}{du} = \frac{1}{T}$$

i.e.
$$\frac{4}{3}\left\{\frac{1}{T}\frac{du}{dT} - \frac{u}{T^2}\right\}\frac{dT}{du} = \frac{1}{T}$$

i.e. $\dfrac{du}{dT} = 4\,\dfrac{u}{T}$, which, integrated, gives $u = aT^4$

where a is a constant.   Now, as we can easily see, the total radiation $K = 2\displaystyle\int_0^\infty \mathbf{K}_\nu d\nu$ is distinguished from the density of radiation u only by a constant factor (see *M. Planck*, Lectures in Radiation (1906), § 22), hence the total radiation is proportional to the fourth power of the absolute temperature and this is the *Stefan-Boltzmann Law*.

11 *W. Wien*, Sitzungsber. d. Akad. d. Wissensch. Berlin, 9 Feb. 1893, p. 55; Wiedem. Ann. 52, 132 (1894). Cf. also *Max Abraham*, Theorie der Elektrizität II, § 43 (1914); *M. Planck*, Vorlesungen über die Theorie der Wärmestrahlung (Leipzig 1906), pp. 68 *et seq.*; *W. Westphal*, Verhandl. d. deutsch. physikal. Ges. 1914, p. 93; *H. A. Lorentz*, Akad. d. Wissensch. Amsterdam, 18 May 1901, p. 607.

12 Formula (4) of the text (*Wien's* Law of Displacement) may be obtained by means of a simple dimensional calculation, as *L. Hopf* recently showed in the "Naturwissenschaften" (8, 109, 110 (1920)). We assume that $\mathbf{K}_\nu$ depends only on ν, T, and the velocity of light c. The dimensions of $\mathbf{K}_\nu$ are obtained from the fact that, according to (1),

$$2\pi \mathbf{K}_\nu d\nu = \frac{\text{energy}}{\text{surface} \times \text{time}}.$$

From this it follows that

$$[\mathbf{K}_\nu] = [mt^{-3}].$$

If we set

$$\mathbf{K}_\nu = \text{const.} \cdot \nu^x \cdot T^y \cdot c^z$$

then, remembering that T has the dimensions of energy, we get

$$[mt^{-2}] = \text{const.} \, [t^{-x} \cdot m^y \cdot l^{2y} \cdot t^{-2y} \cdot l^z \cdot t^{-z}]$$

$$= \text{const.} \, [m^y \cdot l^{2y+z} \cdot t^{-x-2y-z}]$$

Hence $x = 2$; $y = 1$; $z = -2$

which gives us, $\mathbf{K}_\nu = \text{const.} \cdot \dfrac{\nu^2}{c^2} \cdot T$.

This relation is not, however, as we shall see, generally valid. In fact, it would give no finite value for $K = 2\displaystyle\int_0^\infty \mathbf{K}_\nu d\nu$. But, according to the *Stefan-Boltzmann Law* (3), $K = \gamma \cdot T^4$. Hence the constant of $\mathbf{K}_\nu$ may still depend on a dimensionless combination of the four variables γ, ν, T, c. If, therefore, we set const. $= f(\nu^\xi T^\eta c^\zeta \gamma^\omega)$ then the argument of the function f must have the dimension 0. If, further, we remember that

$$[\gamma] = \left[\frac{K}{T^4}\right] = \left[\frac{2\displaystyle\int_0^\infty \mathbf{K}_\nu d\nu}{T^4}\right] = \left[\frac{\frac{\text{energy}}{\text{surface} \times \text{time}}}{\text{energy}^4}\right] = [m^{-3}l^{-8}t^5],$$

it then follows that

$$[\nu^\xi T^\eta c^\zeta \gamma^\omega] = [t^{-\xi} . m^\eta . l^{2\eta} . t^{-2\eta} . l^\zeta . t^{-\zeta} . m^{-3\omega} . l^{-3\omega} . t^{5\omega}]$$

$$= [t^{-\xi-2\eta-\zeta+5\omega} . l^{2\eta+\zeta-3\omega} . m^{\eta-3\omega}.]$$

$$\therefore \qquad \xi = -3\omega; \qquad \eta = 3\omega; \qquad \zeta = 2\omega.$$

Hence $$\text{const.} = f\left[\left(\frac{T}{\nu}\right)^{3\omega} . c^{2\omega} \cdot \gamma^\omega\right] = \phi\left(\frac{\nu}{T}\right).$$

Therefore $$\mathbf{K}_\nu = \frac{\nu^3}{c^2} T . \phi\left(\frac{\nu}{T}\right) = \frac{\nu^2}{c^3} T . \frac{\nu}{T} . F\left(\frac{\nu}{T}\right)$$

or, finally, $$\mathbf{K}_\nu = \frac{\nu^3}{c^3} F\left(\frac{\nu}{T}\right).$$

[13] If we plot $\mathbf{K}_\nu = \dfrac{\nu^3}{c^2} F\left(\dfrac{\nu}{T}\right)$ as a function of ν, keeping T constant, the maximum of this curve—if one is present—lies at that point at which $\dfrac{\partial \mathbf{K}_\nu}{\partial \nu} = 0$. This gives

$$3F\left(\frac{\nu}{T}\right) + \frac{\nu}{T} . F^1\left(\frac{\nu}{T}\right) = 0$$

where F^1 is the differential coefficient of F with respect to the argument. This equation, in which only $\dfrac{\nu}{T}$ occurs as unknown, gives a definite value for $\dfrac{\nu}{T}$. In other words, for $\nu = \nu_{max}$, it follows that $\dfrac{\nu_{max}}{T} = \text{const.}$

[14] W. *Wien*, Wied. Ann. **58**, 662 (1896).

[15] O. *Lummer* and E. *Pringsheim*, Wied. Ann. **63**, 395 (1897); Drude's Ann. **3**, 159 (1900): Verh. d. deutsch. phys. Ges. **1**, 23 and 215 (1899).

The total radiation emitted per second from 1 cm.[2] in one direction is, by formula (1)

$$S = 2\pi \int_0^\infty \mathbf{K}_\nu d\nu.$$

According to the *Stefan-Boltzmann* Law, S is proportional to T^4, therefore $S = \sigma T^4$. (The constant of proportionality σ is related to the constant γ occurring in (3) by the equation $\sigma = \pi\gamma$.) The absolute measurement of S gave the following values for σ, in chronological order:

$\sigma = 5{\cdot}45 . 10^{-12} \left[\dfrac{\text{watt}}{\text{cm.}^2 \text{deg.}^4}\right]$ according to F. *Kurlbaum* [Wiedem. Ann.

$\qquad$ **65**, 746 (1898); Verhandl. d. deutsch. physikal. Ges. **14**, 576, 792 (1912)].

$= 5{\cdot}58 . 10^{-12}$ $\quad$ „ $\quad$ according to S. *Valentiner* [Ann. d. Phys. **31**, 255 (1910); **39**, 489 (1912)].

$= 5{\cdot}90 . 10^{-12}$ $\quad$ „ $\quad$ according to W. *Gerlach* [Ann. d. Phys. **38**, 1 (1912)].

$$\sigma = 5{\cdot}30 \cdot 10^{-12}\left[\frac{\text{watt}}{\text{cm.}^2\,\text{deg.}^4}\right]$$ according to *E. Bauer* and *M. Moulin* [Soc. Franc. de Phys. Nr. **301**, 2–3 (1909)].

$= 6{\cdot}30 \cdot 10^{-12}$,,   according to *Ch. Féry* [Bull. Soc. Franc. Phys. **4** (1909)].

$= 6{\cdot}51 \cdot 10^{-12}$,,   according to *Ch. Féry* and *M. Drecq* [Journ. de Phys. (5) **1**, 551 (1911)].

$= 5{\cdot}67 \cdot 10^{-12}$,,   according to *G. A. Shakespear* [Proc. Roy. Soc. (A) **86**, 180 (1911)].

$= 5{\cdot}54 \cdot 10^{-12}$,,   according to *W. H. Westphal* [Verhandl. d. deutsch. physikal. Ges. **14**, 987 (1912)].

$= 6{\cdot}05 \cdot 10^{-12}$,,   according to *L. Puccianti* [Cim. (6) **4**, 31 (1912)].

$= 5{\cdot}89 \cdot 10^{-12}$,,   according to *Keene* [Proc. Roy. Soc. (A) **88**, 49 (1913)].

$= 5{\cdot}57 \cdot 10^{-12}$,,   according to *W. H. Westphal* [Verhandl, d. deutsch. physikal. Ges. **15**, 897 (1913)].

$= 5{\cdot}85 \cdot 10^{-12}$,,   according to *W. Gerlach* [Phys. Zeitschr. **17**, 150 (1916)].

As regards *Wien's* Law of Displacement, the relation (5a) was tested and found to be confirmed. From Fig. 1, in which E_λ is plotted as a function of λ for different values of λ, we see clearly how the maximum of the curve becomes displaced towards shorter wave-lengths as the temperature rises.

For the constant on the right-hand side of relation (5a) the measurements gave the following values:

const. $= 0{\cdot}294$ [cm. deg.] according to *O. Lummer* and *E. Pringsheim*

                               [Verhandl. d. deutsch. physikal. Ges. **1**, 23 and 215 (1899)].

$= 0{\cdot}292$,,   according to *F. Paschen* [Drude's Ann. **6**, 657 (1901)].

$= 0{\cdot}2911$,,   according to *Coblentz* [Bull. Bur. of Stand. **10**, 1 (1914)].

16 *O. Lummer* and *E. Pringsheim*, Verhandl. d. deutsch. physikal. Ges. **1**, 215 (1899).

17 *F. Paschen*, Berliner Ber. 1899, pp. 405, 959.

18 *M. Planck*, Absorption und Emission elektr. Wellen durch Resonanz. Sitzungsber. d. Berl. Akad. d. Wiss. 21 March 1895, pp. 289–301; Wiedem. Ann. **57**, 1–14 (1896).—Über elektr. Schwingungen, welche durch Resonanz erregt und durch Strahlung gedämpft werden. Sitzungsber. d. Berl. Akad. d. Wiss. 20 Febr. 1896, pp. 151–170; Wiedem. Ann. **60**, 577–599 (1897).—Über irreversible Strahlungsvorgänge. (1. Mitteilung.) Sitzungsber. d. Berl. Akad. d. Wiss., 4 Febr. 1897, pp. 57–68. (2. Mitteilung) *ibid.*, 8 July 1897, pp. 715–717. (3. Mitteilung) *ibid.*, 16 Dec. 1897, pp. 1122–1145. (4 Mitteilung) *ibid.*, 7 July 1898, pp. 449–476. (5. Mitteilung) *ibid.*, 18 May 1899, pp. 440–480. (Supplement.) *ibid.*, 9 May 1901, pp. 544–555;

Drudes Ann. **1**, 69–122 (1900). (Supplement.) Drudes Ann. **6**, 818–831 (1901).—Entropie und Temperatur strahlender Wärme. Drudes Ann. **1**, 719–737 (1900).

19 In place of the mean value, with respect to *time*, of the energy of a *single* oscillator, we may use the *spatial* mean value of the momentary energy of a whole *system consisting of very many oscillators.*

20 In this second, more difficult part of the calculation, *Planck* takes his stand upon the second law of thermodynamics, and seeks, from this view, to determine a phase-quantity S of the oscillator, which possesses the well-known property of the entropy, that it increases in all irreversible processes. He arrived at the solution:

$$S = -\frac{\overline{U}}{\beta\nu}\left\{\log_e\left(\frac{\overline{U}}{a\nu}\right) - 1\right\}$$

This function possessed, as *Planck* showed, the required property of entropy, but it was not the only function with this property. And in fact it appeared later, that in the deduction of the above expression, a readily suggested but unjustified supposition had been made. The expression given in the text, formula (8), for the mean energy $\overline{U}$ follows from S by applying the second law in the form:

$$dS = \frac{d\overline{U}}{T}\ \text{or}\ \frac{dS}{d\overline{U}} = \frac{1}{T}.$$

21 *O. Lummer* and *E. Pringsheim*, Verhandl. d. deutsch. physikal. Ges. 1900, p. 163.

22 *M. Planck*, Verh. d. deutsch. phys. Ges. 1900, p. 237. It is of historic interest to note that *Planck* had already, in a somewhat earlier paper (Verh. d. deutsch. phys. Ges. 1900, p. 202), arrived at the true law of radiation by a purely formal alteration of *Wien's* formula, which was not further explained. Cf. also Ann. d. Phys. **4**, 553 (1901); **4**, 564 (1901); **6**, 818 (1901); **9**, 629 (1902).

23 Let N oscillators be present. Let the total energy to be divided among them be $U_N = N\overline{U}$. The "state" or phase of the oscillator-system, the probability of which is to be calculated, is then defined by the fact that N oscillators possess the energy U_N. We divide U_N into P energy elements ϵ, so that

$$U_N = N \cdot \overline{U} = P\epsilon.$$

The number of possible ways of distributing P balls among N boxes is, however,

$$W = \frac{(N + P - 1)!}{(N - 1)!\,P!}.$$

This is therefore the probability of the state, which corresponds to the distribution of P energy elements among N oscillators. *P. Ehrenfest* and *H. Kamerlingh-Onnes* give a very simple deduction of this formula in Ann. d. Phys. **46**, 1021 (1915).

The rule mentioned in the text, which is due to *Boltzmann*, states that the entropy S_N of the oscillator system is connected with the probability W by the fundamental relation

$$S_N = k \log W$$

where k is a constant.

In this theorem of *Boltzmann* the following law of the growth of entropy (second law of thermodynamics) is contained: if a system passes from an improbable condition into a more probable one, then by this transition W, and therefore the entropy S, increases. If we here insert the value of W, and, since N and P are very large numbers, use *Stirling's* approximation formula

$$\log_e (N!) = N(\log_e N - 1)$$

then, if we set for P, $N\,\dfrac{\bar{U}}{\epsilon}$, we get by an easy calculation

$$S_N = kN\left\{\left(1 + \frac{\bar{U}}{\epsilon}\right)\log\left(1 + \frac{\bar{U}}{\epsilon}\right) - \frac{\bar{U}}{\epsilon}\log\left(\frac{\bar{U}}{\epsilon}\right)\right\}$$

and hence the entropy S of *one* oscillator becomes:

$$S = \frac{S_N}{N} = k\left\{\left(1 + \frac{\bar{U}}{\epsilon}\right)\log\left(1 + \frac{\bar{U}}{\epsilon}\right) - \frac{\bar{U}}{\epsilon}\log\left(\frac{\bar{U}}{\epsilon}\right)\right\}$$

But according to the Second Law (see note 20)

$$\frac{dS}{d\bar{U}} = \frac{1}{T}.$$

If we carry out the differentiation on the left-hand side, and solve the resulting relation between $\bar{U}$, T, and ϵ, with respect to $\bar{U}$, we get the expression (9) of the text.

24 Cf. the paper by *Ehrenfest* and *Kamerlingh-Onnes* cited in the previous note.

25 This law is essentially identical with *Boltzmann's* H-Theorem. Cf. L. *Boltzmann*, Vorlesungen über Gastheorie Bd. I, p. 38 (1896); Sitzungsber. d. Wiener Akad. d. Wiss. (II) 76, 373 (1877). Cf. also P. *Ehrenfest*, Phys. Zeitschr. 15, 657 (1914).

26 H. *Rubens* and F. *Kurlbaum*, Sitzungsber. d. Berl. Akad. d. Wiss. 1900, p. 929; Ann. d. Phys. 4, 649 (1901).

27 F. *Paschen*, Ann. d. Phys. 4, 277 (1901).

28 L. *Holborn* and S. *Valentiner*, Ann. d. Phys. 22, 1 (1907); *Coblentz*, Physical Review, 31, 317 (1910); E. *Baisch*, Ann. d. Phys. 35, 543 (1911); E. *Warburg*, G. *Leithäuser*, E. *Hupka* and C. *Müller*, Ann. d. Phys. 40, 609 (1913); E. *Warburg* and C. *Müller*, Ann. d. Phys. 48, 410 (1915).

29 W. *Nernst* and Th. *Wulf*, Ber. d. deutsch. phys. Ges. 21, 294 (1919).

30 *Lord Rayleigh*, Phil. Mag. 49, 539 (1900).

31 The " *Stefan-Boltzmann* constant of total radiation " σ, introduced in note 15, has therefore the value

$$\sigma = \frac{2\pi^5 k^4}{15c^2 h^3}.$$

[32] In order to determine the constants h and k which occur in the radiation formula, we can, instead of using the equation : $\lambda_{max} . T =$ const., compare other relations with the measurement of the total radiation. For example, we can proceed as follows : At a constant temperature T we measure the ratio of the intensity of radiation for two different wave-lengths λ_1 and λ_2 (isothermal method). Now this ratio is, according to (15)

$$\frac{E_{\lambda_1}}{E_{\lambda_2}} = \left(\frac{\lambda_2}{\lambda_1}\right)^5 \cdot \frac{e^{\frac{C}{\lambda_2 T}} - 1}{e^{\frac{C}{\lambda_1 T}} - 1}, \text{ where } C = \frac{hc}{k}.$$

From this relation, since everything excepting C is known, C, that is, $\frac{h}{k}$ may be calculated. Another method is the following : we measure for a fixed wave-length λ the ratio of the intensity of radiation at two different temperatures T_1 and T_2 (isochromatic method). Then it follows that

$$\frac{E^{(T_1)}_\lambda}{E^{(T_2)}_\lambda} = \frac{e^{\frac{C}{\lambda T_2}} - 1}{e^{\frac{C}{\lambda T_1}} - 1}.$$

This is a relation from which C, that is, $\frac{h}{k}$ can again be calculated.

With the help of these methods, the researches, for example, of *Warburg* and his co-workers cited in note 28 have yielded values for $C = \frac{ch}{k}$ which lie in close proximity to $C = 1\cdot430$. This value was taken by *Nernst* and *Wulf* (see note 29) for their critical investigation.

For the constant of *Wien's* Law of Displacement in the form $\lambda_{max} . T = b$ we would accordingly get from (16) :

$$b = \frac{C}{4\cdot9651} = 0\cdot288$$

a value smaller, therefore, than that given by direct measurement (see note 15). Whether *Warburg's* value, $C = 1\cdot430$, or the measured values of $b(> 0\cdot29)$ or both, are seriously affected by experimental error, or whether after all—as *Nernst* and *Wulf* maintain—*Planck's* formula is not right, must be left for the future to decide.

[33] *M. Planck*, Ann. d. Phys. **4**, 553 (1901).

[34] If we apply *Boltzmann's* relation $S = k \log W$ (quoted in note 15), which connects the entropy S with the probability of state W, to one gramme-molecule of an ideal gas, then by calculating the probability of a certain state, i.e. a certain distribution of velocities among the molecules, we arrive at the following value for the entropy of the gas

$$S = kN(\tfrac{3}{2} \log_e U + \log V) + \text{const.}$$

(Cf., for example, M. *Planck*, Lectures on the Theory of Radiation (1906), § 143.) Here N is the number of molecules in a gramme-molecule (*Avogadro's* number), U the energy, V the volume of the gas. Now, according to the Second Law of Thermodynamics,

$$dS = \frac{dU + pdV}{T}$$

must be a complete differential, where p and T denote pressure and temperature of the gas. Hence the relation

$$\left(\frac{\partial S}{\partial V}\right)_U = \frac{p}{T}$$

must hold. This gives

$$\frac{kN}{V} = \frac{p}{T}, \text{ i.e. } p = \frac{kNT}{V}.$$

If we compare this with the equation of state of an ideal gas in thermo-dynamics, $p = \dfrac{RT}{V}$, we get for the absolute gas constant R the value

$$R = kN$$

from which formula (19) of the text follows.

35 *M. Planck*, Ann. d. Phys. **4**, 564-566 (1901).

36 Compare, for example, the table of the values of *Avogadro's* number given in the report of *J. Perrin* at the Solvay Congress in Brussels (1911). [*A. Eucken*, Die Theorie der Strahlung und der Quanten. Abhandlungen der Bunsen-Gesellschaft Nr. 7, Wilh. Knapp, Halle 1914.]

37 *R. A. Millikan*, Phil. Mag. (6) **34**, 13 (1917).

38 *Ibid.*, from the values given by *Millikan* for the electronic charge $e = 4.774 \times 10^{-10}$ (electrostatic units) and from the electrochemical constant $F = 969.4 . 2.999 . 10^{10}$ electrostatic units, there follows for *Avogadro's* number the value $N = 6.0617 . 10^{23}$.

39 Cf., for example, *W. Gibbs'* Elements of Statistical Mechanics, Chapter V.

40 The term "mean value" may be taken as referring to time or to space. If we select a definite atom, and follow it a long time upon its zig-zag path, and from the mean of the values which its kinetic energy assumes in the course of time, we get the "time-mean." If, on the other hand, we select a large number of identical atoms of the gas at a particular instant and again form the mean of the values of the kinetic energies which these atoms possess at the instant in question, we get the "space-mean."

41 If x is the elongation of the oscillator (electron) vibrating with the natural frequency, then $x = A \sin(2\pi\nu t)$, where A is the amplitude and t the time; the mean kinetic energy becomes

$$\overline{L} = \tfrac{1}{2}m\overline{\left(\frac{dx}{dt}\right)^2} = \tfrac{1}{2}m(A \cdot 2\pi\nu)^2 . \overline{\cos^2(2\pi\nu t)} = \tfrac{1}{4}m(2\pi\nu A)^2.$$

The mean potential energy is:

$$\overline{V} = \tfrac{1}{2}m(2\pi\nu)^2\overline{x^2} = \tfrac{1}{2}m(2\pi\nu A)^2 . \overline{\sin^2(2\pi\nu t)} = \tfrac{1}{4}m(2\pi\nu A)^2.$$

Hence, as stated, $\overline{L} = \overline{V}$: i.e. the mean kinetic energy = the mean potential energy.

42 *J. H. Jeans*, Phil. Mag. **10**, 91 (1905).

43 *H. A. Lorentz*, Proc. Kon. Akad. v. Wet., Amsterdam 1903, p. 666. —The theory of electrons (Teubner, Leipzig 1909), Ch. II.

44 *A. Einstein* and *L. Hopf*, Ann. d. Phys. **33**, 1105 (1910).

45 *A. D. Fokker*, Ann. d. Phys. **43**, 810 (1914).

46 *M. Planck*, Ber. d. Berl. Akad. d. Wiss., 8 July 1915, p. 512.

47 *H. A. Lorentz*. Die Theorie d. Strahlung u. d. Quanten ; Abhandlungen der Deutschen Bunsen-Gesellschaft. Nr. 7. v. *A. Eucken*. Halle, W. Knapp 1914 pp. 10 *et seq.*

48 By a suitable modification of classical statistics in the sense of the quantum theory, we can obtain the expression (9) for the mean energy of an oscillator in the following manner which is worthy of notice. Let a number N of similar oscillators with the most varied values for the energy be given. We require to find how great is the probability w, that an oscillator possess a certain energy value U; or, otherwise expressed, how many of the N oscillators possess the energy U. In order to answer this question, we find it best to take first of all the standpoint of *Gibbs'* statistical mechanics, that is, of "classical" statistics. In place of the special case in question, namely, that of the linear oscillator, let us consider at once quite generally a system of f degrees of freedom, and characterise it by f generalised co-ordinates $q_1 q_2 \ldots q_f$ and by the corresponding impulses or momenta $p_1 p_2 \ldots p_f$. (Here, the impulse p_i is thus defined: form the kinetic energy of the system as a function of the generalised velocities $\dot{q}_i = \dfrac{dq^i}{dt}$, then $p_i = \dfrac{\partial L}{\partial \dot{q}_i}$.) In particular, the linear oscillator (vibrating electron) will be described by a co-ordinate q, namely, the elongation of the electron, and the impulse $p = m\dfrac{dq}{dt}$. In general, therefore, $2f$ quantities are necessary in order to define completely the momentary state of a system. Hence we can represent this momentary state by a point ("phase-point") in the $2f$-dimensional space in which $q_1 \ldots p_f$ (of the "phase-space") are co-ordinates.

We now consider a number N of similar systems of this kind, which are in thermodynamic equilibrium with a very large reservoir at the temperature T. Then the probability that the co-ordinates and impulses lie in the small intervals $q_1 \ldots q_1 + dq_1$, etc., and $p_1 \ldots p_1 + dp_1$, etc., that is, that the "phase-point" of the system lie in the element $d\Omega = dq_1 dq_2 \ldots dq_f, dp_1 dp_2 \ldots dp_f$ of the phase-space is, according to *Gibbs*,

$$w = \frac{e^{-\frac{E}{kT}}d\Omega}{\int e^{-\frac{E}{kT}}d\Omega}.$$

Here E is the energy of the system, and k is the constant defined in (19). The integration in the denominator is to be taken over all possible values of the $2f$ quantities $q_1 \ldots p_f$, or, as we may say, over all possible "phases," or over the whole region of the phase-space concerned.

Among the N systems there are then Nw, whose phase-points lie in the element $d\Omega$ of the phase-space. *This is therefore a "distribution" of the N systems over the phase-space.* This distribution is called *Canonical;* it represents a generalisation of *Maxwell's* familiar law of distribution of velocities which may be deduced from it by special-ising it for the case of the gas atom, that is, by setting $f = 3$.

The sum of all probabilities is naturally 1. Indeed, it is at once clear that

$$\Sigma w = \frac{\int e^{-\frac{E}{kT}} d\Omega}{\int e^{-\frac{E}{kT}} d\Omega} = 1.$$

For the mean value of the energy E we get

$$\overline{E} = \Sigma E w = \frac{\int E e^{-\frac{E}{kT}} d\Omega}{\int e^{-\frac{E}{kT}} d\Omega}.$$

If we apply this equation to the linear oscillator we get

$$\overline{U} = \frac{\iint U e^{-\frac{U}{kT}} dq dp}{\iint e^{-\frac{U}{kT}} dq dp}.$$

Now,
$$U = \frac{m}{2}\left(\frac{dq}{dt}\right)^2 + \frac{m}{2} \cdot (2\pi\nu)^2 q^2$$

i.e.
$$U = \frac{p^2}{2m} + 2\pi^2\nu^2 m q^2.$$

If we introduce the auxiliary variables ξ and η, defined by

$$\begin{cases} \xi = \pi\nu q \sqrt{2m} \\ \eta = \frac{p}{\sqrt{2m}}, \text{ and hence } dq dp = \frac{1}{\pi\nu} d\xi d\eta \end{cases}$$

we get

$$U = \xi^2 + \eta^2$$

and, therefore, it suggests itself to us to write

$$\begin{cases} \xi = \sqrt{U} \cos\phi \\ \eta = \sqrt{U} \sin\phi; \end{cases}$$

where ϕ is a parametric angle. If we interpret ξ and η as Cartesian co-ordinates of a point in the plane, then $\sqrt{\overline{U}}$ and ϕ are the polar co-ordinates of this point. The element of surface $d\xi d\eta$ is written in polar co-ordinates, as we know, thus

$$d\xi d\eta = \sqrt{U}(d\sqrt{U})d\phi = \tfrac{1}{2}dUd\phi$$

hence

$$dqdp = \frac{1}{\pi\nu}d\xi d\eta = \frac{1}{2\pi\nu}dUd\phi.$$

Hence

$$U = \frac{\displaystyle\int_{U=0}^{\infty}\int_{\phi=0}^{2\pi} Ue^{-\frac{U}{kT}}dUd\phi}{\displaystyle\int_{U=0}^{\infty}\int_{\phi=0}^{2\pi} e^{-\frac{U}{kT}}dUd\phi} = \frac{\displaystyle\int_{0}^{\infty} Ue^{-\frac{U}{kT}}dU}{\displaystyle\int_{0}^{\infty} e^{-\frac{U}{kT}}dU} = kT$$

in agreement with (24). This is the standpoint of *classical statistics*.

The *quantum statistics* of the oscillator may be immediately deduced from this, if we elaborate the canonical law of distribution

$$w = \frac{e^{-\frac{U}{kT}}dqdp}{\displaystyle\int e^{-\frac{U}{kT}}dqdp}$$

in a suitable manner

If we here again introduce $dqdp = \dfrac{1}{2\pi\nu}dUd\phi$, and integrate with respect to ϕ, we get

$$w_U = \frac{e^{-\frac{U}{kT}}dU}{\displaystyle\int e^{-\frac{U}{kT}}dU}$$

as the probability that the energy of the oscillators lies between U and $U + dU$.

Now the quantum theory demands that the energy U shall assume only the discrete values U_0, U_1, U_2, ... U_n. The transition may best be effected by laying down the condition: E shall only be able to assume the values contained in the narrow intervals between U_0 and $U_0 + \alpha$, U_1 and $U_1 + \alpha$, and generally U_n and $U_n + \alpha$. Then $dU = \alpha$, and the integral in the denominator changes into a sum. Thus it follows that

$$w_{U^n} = w_n = \frac{e^{-\frac{U_n}{kT}} \cdot \alpha}{\displaystyle\sum_{n} e^{-\frac{U_n}{kT}} \cdot \alpha} = \frac{e^{-\frac{U_n}{kT}}}{\displaystyle\sum_{n} e^{-\frac{U_n}{kT}}}$$

thus α is eliminated; if we now proceed to the limit $\alpha = 0$, w remains unaltered. *Hence w_n is the canonical distribution function generalised for quantum conditions*, and hence, among N oscillators, Nw_n have an energy of the value U_n.

We now get for the mean energy

$$\overline{U} = \sum_n U_n w_n = \frac{\sum\limits_n U_n e^{-\frac{U_n}{kT}}}{\sum\limits_n e^{-\frac{U_n}{kT}}}$$

Now, according to the first form of the quantum theory,

$$U_n = n\epsilon = nh\nu \qquad (n = 0, 1, 2, 3 \ldots \infty).$$

Therefore

$$\overline{U} = \frac{\epsilon \sum\limits_0^\infty n e^{-\frac{n\epsilon}{kT}}}{\sum\limits_0^\infty e^{-\frac{n\epsilon}{kT}}} = \epsilon \frac{S_1}{S_2}.$$

If we set $\dfrac{\epsilon}{kT}$, for convenience, $= x$, then

$$S_1 = \sum_0^\infty n e^{-nx}; \quad S_2 = \sum_0^\infty e^{-nx} = \frac{1}{1 - e^{-x}}.$$

Further,

$$-\frac{dS_2}{dx} = \sum_0^\infty n e^{-nx} = S_1 = \frac{e^{-x}}{(1 - e^{-x})^2}$$

from which we get

$$\overline{U} = \epsilon \frac{e^{-x}}{1 - e^{-x}} = \frac{\epsilon}{e^x - 1} = \frac{\epsilon}{e^{\frac{\epsilon}{kT}} - 1}$$

in agreement with (9)

The canonical distribution may be still further generalised by the introduction of certain "weight factors," which are intended to express the fact that the individual quantum states of the system considered have, *a priori*, different probabilities. This happens, for example, if each quantum state may be realised in different ways, and if the number of these possibilities of realisation is different for the different quantum states. Then, the different states will have different "weights," and a "weight factor" p_n has to be included in the exper mental function $e^{-\frac{U_n}{kT}}$ so that the canonical distribution function assumes the form

$$w_n = \frac{p_n e^{-\frac{U_n}{kT}}}{\sum p_n e^{-\frac{U_n}{kT}}} = C \cdot p_n e^{-\frac{U_n}{kT}}.$$

Here C depends on the temperature; p_n, on the other hand, does not.

49 *A. Einstein*, Ann. d. Phys. 17, 132 (1905) ; 20, 199 (1906) ; Verhandl. d. deutsch. physikal. Ges. 11, 482 (1909) ; Bericht Einstein auf dem Solvay-Kongress in Brussels 1911 ; cf. *A. Eucken*, Die Theorie der Strahlung und der Quanten ; Abhandl. d. deutsch. Bunsen-Gesellschaft, Nr. 7 (Halle, W. Knapp 1914), pp. 330 *et seq.* Cf. also *W. Wien*, Vorlesungen über neuere Probleme der theoretischen Physik (Teubner, Leipzig and Berlin 1913), 4. Vorlesung. *H. A. Lorentz*, Les théories statistiques en thermodynamique (Teubner, Leipzig and Berlin 1916), §§ 42 *et seq.*

50 *A. Einstein*, Ann. d. Phys. 17, 132 (1905).

51 *A. Einstein*, Phys. Zeitschr. 10, 185 (1909).

52 This formula may be deduced as follows : Firstly, from $\epsilon = E - \bar{E}$ the frequently used relation

$$\overline{\epsilon^2} = \overline{E^2} - \overline{2\bar{E} \cdot E} + (\bar{E})^2 = \overline{E^2} - (\bar{E})^2$$

follows. In order now to calculate the two quantities $\overline{E^2}$ (mean of the squares of the energy) and $(\bar{E})^2$ (square of the mean energy), which are known to differ from each other in general, we do best to take the standpoint of *Gibbs'* statistical mechanics (see note 48). According to this, the probability that the co-ordinates and impulses lie in the small intervals $q_1 \ldots q_1 + dq_1$, etc., $p_1 \ldots p_1 + dp_1$, etc, that is, that the "phase-point" lies in the element $dq_1 dq_2 \ldots dq_f dp_1 dp_2 \ldots dp_f = d\Omega$ of the "phase-space" :

$$w = \frac{e^{-\frac{E}{kT}} d\Omega}{\int e^{-\frac{E}{kT}} d\Omega}.$$

Then the mean of the energy follows in the usual way :

$$\bar{E} = \frac{\int E e^{-\frac{E}{kT}} d\Omega}{\int e^{-\frac{E}{kT}} d\Omega}.$$

Likewise,

$$\overline{E^2} = \frac{\int E^2 e^{-\frac{E}{kT}} d\Omega}{\int e^{-\frac{E}{kT}} d\Omega}.$$

We then form

$$\frac{d\bar{E}}{dT} = \frac{\int e^{-\frac{E}{kT}} d\Omega \cdot \frac{1}{kT^2} \int E^2 e^{-\frac{E}{kT}} d\Omega - \frac{1}{kT^2} \left(\int E e^{-\frac{E}{kT}} d\Omega \right)^2}{\left(\int e^{-\frac{E}{kT}} d\Omega \right)^2}$$

$$= \frac{1}{kT^2} [\overline{E^2} - (\bar{E})^2] = \frac{\overline{\epsilon^2}}{kT^2}.$$

Therefore,

$$\overline{\epsilon^2} = kT^2 \frac{d\overline{E}}{dT}$$

We also arrive at the same formula, if instead of the classical canonical distribution function, we start from the quantum distribution function

$$w_n = \frac{p_n e^{-\frac{E_n}{kT}}}{\sum_n p_n e^{-\frac{E_n}{kT}}}.$$

53 The mean energy of radiation of frequency ν in the volume v is $\overline{E} = vu_\nu d\nu$, where the monochromatic density of radiation is

$$u_\nu = \frac{8\pi \mathbf{K}_\nu}{c} = \frac{8\pi h\nu^3}{c^3} \cdot \frac{1}{e^{\frac{h\nu}{kT}} - 1}$$

if *Planck's* Law is taken as the basis. (Cf., for example, M. *Planck*, Lectures on the Theory of Radiation, Engl. Transl.)

According to formula (28) deduced in the previous note, it therefore follows that

$$\overline{\epsilon^2} = kT^2 \frac{d\overline{E}}{dT} = kT^2 v d\nu \frac{du_\nu}{dT} = \frac{8\pi h^2 \nu^4 v d\nu e^{\frac{h\nu}{kT}}}{c^3 \left(e^{\frac{h\nu}{kT}} - 1\right)^2}.$$

If we eliminate T on the right-hand side by substituting for $e^{\frac{h\nu}{kT}}$ its value $1 + \frac{8\pi h\nu^3}{c^3 u_\nu}$, it follows that

$$\overline{\epsilon^2} = u_\nu v d\nu \cdot h\nu + \frac{c^3 u_\nu^2 v d\nu}{8\pi \nu^2}$$

i.e.

$$\overline{\epsilon^2} = \overline{E} \cdot h\nu + \frac{c^3 (\overline{E})^2}{8\pi \nu^2 v d\nu}.$$

The second term on the right is required by the Undulatory Theory for at each point of the volume v the most varied trains of waves of radiation cross one another's paths with every possible amplitude and phase. The interference of all these waves thus generates at the point considered an intensity, which varies continually, and hence the energy of the volume v also varies. If we calculate the mean of the square of the energy, i.e. $\overline{\epsilon^2}$, we find precisely the second term of the above formula. (Cf., for example, H. A. *Lorentz*, Les théories statistiques en thermodynamique (Teubner, Leipzig and Berlin), 1916, pp. 114 *et seq.*)

The first term is *not*, however, explained by the classical undulatory theory. On the other hand, it becomes endowed with meaning if we suppose that the radiant energy consists of a certain whole number

(n) of finite energy complexes of the value $h\nu$. For then $E = n \cdot h\nu$, and therefore $\bar{E} = \bar{n} \cdot h\nu$, where $\bar{n}$ is the mean about which the number n varies. If $\delta = n - \bar{n}$ be the variation of the number n, then it follows that $\epsilon = E - \bar{E} = \delta h\nu$, where $\bar{\epsilon^2} = \bar{\delta^2} \cdot h^2\nu^2$. But, according to a well-known law of statistics, $\bar{\delta^2} = \bar{n}$. (Cf., for example, H. A. *Lorentz*, *loc. cit.*, §§ 26 and 27.) Hence $\bar{\epsilon^2} = \bar{n}h^2\nu^2 = \bar{E} \cdot h\nu$. This is exactly the first term in the above formula.

[54] *A. Einstein*, Ann. d. Phys. **17**, 144 (1905).

[55] *J. J. Thomson*, Conduction of Electricity through Gases.

[56] *A. Einstein*, Ann. d. Phys. **17**, 147 (1905).

[57] Cf. R. *Pohl* and P. *Pringsheim*, Die lichtelektrischen Erscheinungen. Sammlung Vieweg Heft 1 (Braunschweig 1914).

[58] *A. Einstein*, Ann. d. Phys. **17**, 145 (1905).

[59] *R. A. Millikan*, Phys. Zeitschr. **17**, 217 (1916).

[60] According to *Pohl* and *Pringsheim*, we have to distinguish between the *normal* and the *selective* photo-effect: in the case of the normal effect the number of electrons torn off (per calorie of the light-energy absorbed) is independent of the orientation of the electrical vector of the light-wave, and increases, starting from an upper limit of the wave-length, in general uniformly as the wave-length decreases. In the case of the selective effect, on the other hand, which only appears when the electrical vector of the light-wave possesses a component vertical to the metallic surface, the number of electrons torn off (per calorie of light-energy absorbed) shows a decided maximum at a definite wave-length.

[61] *Ch. Barkla*, Phil. Mag. **7**, 543, 812; **15**, 218. Jahrb. d. Radioaktivität u. Elektronik, 5, p. 239, 1908.—*Ch. Barkla* and *Sadler*, Phil. Mag. **17**, 739.—*Ch. Barkla*, Jahrb. d. Radioaktivität u. Elektronik, 1910, p. 12.—*M. de Broglie*, C. R. 25 May and 15 June 1914, p. 1785.—*Ch. Barkla*, Phil. Mag. **16**, 550.—*E. Wagner*. Ann. d. Phys. **46**, 868 (1915); Sitzungsber. d. bayer. Akad. 1916, p. 33.

[62] *D. L. Webster*, Proc. Americ. Acad. **2**, 90 (1916); Physic. Review, **7**, 587 (1916).

[63] *E. Wagner*, Ann. d. Phys. **46**, 868 (1915).

[64] Cf., for example, *E. Wagner*, Phys. Zeitschr. **18**, 443 (1917). The value that *Wagner* calculates for h is: $h = 6 \cdot 62 \cdot 10^{-27}$.

[65] *W. Duane* and *F. L. Hunt*, Physic. Review, **6**, 166 (1915).

[66] *A. W. Hull* and *M. Rice*, Proc. Americ. Acad. **2**, 265 (1916).

[67] *E. Wagner*, Phys. Zeitschr. **18**, 440 *et seq.* (1917); Ann. d. Phys. **57**, 401 (1918).

[68] *F. Dessauer* and *E. Back*, Ber. d. deutsch. physikal. Ges. **21**, 168 (1919).

[69] *J. Franck* and *G. Hertz*, Verhandl. d. deutsch. physikal. Ges. **16**, 512 (1914).

[70] The critical potential measured by *Franck* and *Hertz* amounted to $V = 4 \cdot 9$ volts $= \dfrac{4 \cdot 9}{300}$ electrostatic units, and therefore the critical energy of the electron is

$$eV = \frac{4 \cdot 774 \cdot 10^{-10} \cdot 4 \cdot 9}{300}.$$

The wave-length λ of the mercury line emitted is

$$\lambda = 2536\overset{\circ}{A} = 2\cdot536 \,.\, 10^{-5}.$$

Hence we must get

$$eV = h\frac{c}{\lambda}, \text{ i.e. } h = \frac{eV\lambda}{c} = \frac{4\cdot774 \,.\, 10^{-10} \,.\, 4\cdot9 \,.\, 2\cdot536 \,.\, 10^{-5}}{3\cdot10^2 \,.\, 3\cdot10^{10}}$$
$$= 6\cdot59 \,.\, 10^{-27}$$

and this is in good agreement with the results of other measurements.

[71] Cf., for example, J. *Stark*, Prinzipien der Atomdynamik II. (S. Hirzel, Leipzig 1911), Chs. IV and V.

[72] J. *Stark*, Ber. d. deutsch. phys. Ges. **10**, 713 (1908); Phys. Zeitschr. **8**, 913 (1907); **9**, 767 (1908).

Canal-rays are positively charged particles of matter, which move in a vacuum tube in the direction : anode to cathode; the latter is pierced with holes through which the canal-rays pass into the space behind the cathode. If we generate such canal-rays in a vacuum tube filled with hydrogen, we find that the series lines of hydrogen are emitted. Now, if we observe this emission spectroscopically "from the front," that is, so that the canal-rays are moving towards the observer, we see, firstly, at its usual place in the spectrum, the sharp series line (line of rest, "intensity of rest"); secondly, we see displaced towards the violet, a broadened strip (line of motion, "intensity of motion" or "dynamic intensity"). These lines represent the series line emitted by the moving canal-ray particles, which is displaced towards the region of higher frequencies on account of the *Doppler* effect. Since the canal-rays do not possess a single uniform velocity, and since particles with all possible velocities occur, the displaced strip is not sharp, but softened and broadened. The "intensity at rest" is therefore emitted when the quickly moving canal particles strike "resting" molecules, i.e. gas-molecules which are moving comparatively slowly and irregularly, *and excite these to emit the series lines*. The "intensity of motion," on the other hand, is excited by *the unidirectionally moving canal particles themselves*, when they hit gas-molecules.

Now, it is very remarkable that the interval between the intensity of rest and that of motion is not filled in, but that the emission of the intensity of motion becomes observable only above a certain velocity. *Stark* interpreted this fact in terms of the light-quantum hypothesis thus: If $\frac{1}{2}mv^2$ is the kinetic energy of a canal-ray particle, and if the fraction $a\frac{1}{2}mv^2 (a > 1)$ is transformed into a light-quantum $h\nu$ upon collision with a gas-molecule, then we must have $h\nu < \frac{a}{2}mv^2$; that is, the spectral line of frequency ν can be generated only by canal-rays, the velocity of which $\geq \sqrt{\dfrac{2h\nu}{am}}$.

The proportionality between the critical velocity and $\sqrt{\nu}$ has been fairly well borne out.

It should be remarked here that J. *Stark* has lately abandoned the theory of light-quanta. (Cf. J. *Stark*, Verh. d. deutsch. physik. Ges. **16**, 304 (1904); **18**, 42 (1916).)

73 J. *Stark*, Phys. Zeitschr. **9**, 85, 356 (1908).—*J. Stark* and *W. Steubing*, Phys. Zeitschr. **9**, 481 (1908).—*J. Stark*, Phys. Zeitschr. **9**, 889 (1908).

In these papers *J. Stark* defends the view that the band-spectra are emitted when a "valency electron" belonging to the atom or molecule is pushed out of its normal position and then returns again to its initial position, counterbalancing the work done in displacement. If the energy of deformation (valency energy) E is changed into a light-quantum, then we must have $h\nu = E$, i.e. $\nu \lessgtr \frac{E}{h}$. All lines of the band must therefore lie below the edge $\nu = \frac{E}{h}$. If the valency energy E is changed by chemical processes, the band-spectrum must be displaced accordingly.

74 *A Einstein*, Ann. d. Phys. **17**, 148 (1905).

75 J. *Stark*, Phys. Zeitschr. **9**, 889 (1908); Ann. d. Phys. **38**, 467 (1912).

The fundamental law of photochemical decomposition enunciated by *Stark* and *Einstein* states: If a molecule dissociates at all owing to the absorption of radiation of frequency ν, then it will absorb an amount of energy $h\nu$ when it dissociates. This energy, therefore, represents the heat of reaction, which will be set free upon recombination of the products of decomposition.

This law was later deduced by *A. Einstein* for the range of validity of *Wien's* Law of Radiation without the assistance of the light-quantum hypothesis, by purely thermodynamical methods. (Cf. Ann. d. Phys. **37**, 832 (1912), and **38**, 881 (1912).)

76 *E. Warburg*, Ber. d. Berl. Akad. d. Wiss. 1911, p. 746; 1913, p. 644; 1914, p. 872; 1915, p. 230; 1916, p. 314; 1918, pp. 300, 1228. Cf. also "Naturwissenschaften," **5**, 489 (1917).

77 *H. A. Lorentz*, Phys. Zeitschr. **11**, 1250 (1910).

78 *M. Planck*, Ber. d. deutsch. physikal. Ges. **13**, 138 (1911); Ann. d. Phys. **37**, 642 (1912).

79 On account of the continuous (classical) absorption, all energy values of the oscillator in an elementary region, say between $n\epsilon$ and $(n + 1)\epsilon$, are equally probable. The mean energy in the nth elementary region is, therefore,

$$\overline{U}_n = \frac{n\epsilon + (n + 1)\epsilon}{2} = (n + \tfrac{1}{2})\epsilon.$$

From the canonical law of distribution extended in the sense of the quantum theory, it then follows that

$$\overline{U} = \frac{\sum_{0}^{\infty}\overline{U}_n . e^{-\frac{\overline{U}_n}{kT}}}{\sum_{0}^{\infty} e^{-\frac{\overline{U}_n}{kT}}} = \frac{e^{-\frac{\epsilon}{kT}}\sum_{0}^{\infty}\left(n\epsilon + \frac{\epsilon}{2}\right). e^{-\frac{n\epsilon}{kT}}}{e^{-\frac{\epsilon}{2kT}}\sum_{0}^{\infty} e^{-\frac{n\epsilon}{kT}}}$$

$$\frac{\sum\limits_{0}^{\infty} n\epsilon e^{-\frac{n\epsilon}{kT}}}{\sum\limits_{0}^{\infty} e^{-\frac{n\epsilon}{kT}}} + \frac{\epsilon}{2} = \frac{\epsilon}{e^{\frac{\epsilon}{kT}} - 1} + \frac{\epsilon}{2}$$

(cf. note 48). If we further set $\epsilon = h\nu$ it follows that

$$\bar{U} = \frac{h\nu}{e^{\frac{h\nu}{kT}} - 1} + \frac{h\nu}{2}.$$

In place of relation (7) of the text we get here

$$\mathbf{K}_\nu = \frac{\nu^2}{c^2}\left(\bar{U} - \frac{h\nu}{2}\right)$$

and this leads to *Planck's* Law of Radiation.

80 *M. Planck*, Sitzungsber. d. Kgl. Preuss. Akad. d. Wiss. 3 April, 1913, p. 350; *ibid.*, 30 July 1914, p. 918; *ibid.*, 8 July 1915, p. 512.

81 *A. Einstein* and *O. Stern*, Ann. d. Phys. **40**, 551 (1913).

82 *W. Nernst*, Verhandl. d. deutsch. physikal. Ges. **18**, 83 (1916).

83 *F. Richarz*, Wiedem. Ann. **52**, 410 (1894).

84 Report by *P. Langevin* at the Solvay Congress in Brussels, 1911. Cf. *A. Eucken*, Die Theorie der Strahlung und der Quanten. Abhandl. d. deutsch. Bunsen-Ges., Nr. 7 (W. Knapp, Halle 1914), pp. 318 *et seq.*

85 *A. Einstein* and *W. J. de Haas*, Verhandl. d. deutsch. physikal. Ges. **17**, 152, 203, 420 (1915).—*A. Einstein ibid.*, 18, 173 (1916).—*W. J. de Haas, ibid.*, **18**, 423 (1916).

86 *E. Beck*, Ann. d. Phys. **60**, 109 (1919).

87 Report by *Planck* at the Solvay Congress in Brussels, 1911. See *A. Eucken*, Die Theorie der Strahlung und der Quanten. Abhandl. d. deutsch. Bunsen-Ges., Nr. 7 (W. Knapp, Halle 1914), p. 77.

88 If q is the elongation of a linearly vibrating electron of mass m (oscillator) and ν its period of oscillation, then the energy of this configuration is

$$U = \frac{m}{2}\left(\frac{dq}{dt}\right)^2 + \frac{m}{2}(2\pi\nu)^2 \cdot q^2.$$

The first term represents the kinetic and the second the potential energy. Now the impulse (the momentum) is $p = m\frac{dq}{dt}$. Therefore, we may write

$$U = \frac{p^2}{2m} + 2\pi^2\nu^2 m q^2$$

i.e.

$$\frac{q^2}{\left(\sqrt{\dfrac{U}{2\pi^2\nu^2 m}}\right)^2} + \frac{p^2}{(\sqrt{2mU})^2} = 1.$$

The curves $U = $ const., that is, those curves in the phase-plane, which correspond to the states of *constant energy* of the oscillator, are therefore ellipses with the semi-axes

$$a = \sqrt{\frac{U}{2\pi^2\nu^2 m}} \quad \text{and} \quad b = \sqrt{2mU}.$$

For a definite value of U we get a completely definite ellipse. The " phase-point " of the oscillators would continually revolve in this ellipse, if the electron, without emitting or absorbing, were to execute pure harmonic oscillations : for then its energy would remain permanently constant. If we allow U to vary continuously, i.e. if we give it other and again other values in continuous succession, we get an unlimited manifold of concentric ellipses.

The quantum theory, as formulated in (30) in the text, selects from this infinite manifold a *discrete set* of ellipses, and distinguishes them as the " quantised " ellipses which correspond to the " characteristic states " of the oscillator. To these belong the " quantum energy-values " U_0, U_1, U_2 . . . U_n.

Now the nth ellipse encloses an area nh. The area of the nth ellipse is, however,

$$F_n = \pi a_n \cdot b_n = \pi\sqrt{\frac{U_n}{2\pi^2\nu^2 m}} \cdot \sqrt{2mU_n} = \frac{U_n}{\nu}$$

hence we must have

$$\frac{U_n}{\nu} = nh \quad \text{i.e.} \quad U_n = nh\nu$$

that is, *in the nth quantum state the oscillator possesses an amount of energy $n\epsilon = nh\nu$.*

89 *A. Sommerfeld*, Phys. Zeitschr. **12**, 1057 (1911).—Report by *A. Sommerfeld* at the Solvay Congress in Brussels, 1911. Cf. *A. Eucken*, Die Theorie der Strahlung und der Quanten. Abhandl. d. deutsch. Bunsen-Ges., Nr. 7 (W. Knapp, Halle 1914), p. 252.

90 Report by *Sommerfeld* at the Solvay Congress, 1911.

91 *A. Sommerfeld* and *P. Debye*, Ann. d. Phys. **41**, 873 (1913).

92 Cf., for example, the recent summary by *E. Schrödinger*, Der Energieinhalt der Festkörper im Lichte der neueren Forschung. Phys. Zeitschr. **20**, 420, 450, 474 (1919). A complete set of references accompanies this account.

93 One gramme-atom of a substance, the atomic weight of which is a, is defined as the quantity a grammes of the substance. For example, one gramme-atom of copper is equal to 63·57 grammes of copper, since 63·57 is the atomic weight of copper. Exactly analogous is the definition of the gramme-molecule (also called " mol "). One gramme-molecule of oxygen is 32 grammes of oxygen, for the molecular weight of oxygen (diatomic) is 32.

If c is the specific heat of a substance of atomic weight a, it signifies that *one gramme* of the substance requires an amount of heat c to raise its temperature by $1°$ C. Hence we must communicate to a gramme-atom of the substance, i.e. to a grammes of it, an amount of heat $C = ca$ in order to raise its temperature by 1° C. C is then called the *atomic heat*.

94 The equality of the mean potential and the mean kinetic energies is true here as in the case of the linear *Planck* oscillator (vibrating electron), cf. note 41. This equality is, in general, always present when the forces which act upon the atoms and restore them to their positions of rest (zero positions) are *linear functions* of the relative displacements of the atoms, that is, when the force is "quasi-elastic," that is, proportional to the displacement from the zero position. Cf. in this connexion L. *Boltzmann*, Wiener Ber. 63 (11), 731 (1871), and *F. Richarz*, Wied. Ann. 67, 702 (1899).

95 *Dulong* and *Petit*, Ann. de chim. et de phys. 10, 395 (1819).

96 The quantity usually obtained by measurement is not the atomic heat at constant volume C_v, but the atomic heat at constant pressure C_p. For this we get values which in general fluctuate about the value 6·4 cal./deg. The calculation of C_v from C_p is based on the thermodynamically deduced formula

$$C_p - C_v = \frac{a^2 V T}{\kappa}$$

where a is the cubical coefficient of thermal expansion, κ the (isothermal) cubical compressibility, and V the atomic volume $= \dfrac{\text{atomic weight}}{\text{density}}$.

97 E.g. we find

for silver	at	0° C.	.	.	.	$C_p = 6\cdot00$
„ aluminium	„	58° C.	.	.	.	$C_p = 5\cdot82$
„ copper	„	17° C.	.	.	.	$C_p = 5\cdot79$
„ lead	„	17° C.	.	.	.	$C_p = 6\cdot33$
„ iodine	„	25° C.	.	.	.	$C_p = 6\cdot64$
„ zinc	„	17° C.	.	.	.	$C_p = 6\cdot03$

98 *F. H. Weber*, Poggend. Ann. 147, 311 (1872); 154, 367, 553 (1875).

99 As a possible way out, the "agglomeration hypothesis," supported by *F. Richarz* [Marburger Ber. 1904, p. 1], *C. Benedicks* [Ann. d. Phys. 42, 133 (1913)] and others, has been put forward. According to this, as the temperature falls the number of degrees of freedom of the system diminishes by "freezing-in," as it were, in that certain linkages become completely rigid. According to this, however, the compressibility should decrease greatly as the temperature falls, which, according to *E. Grüneisen's* measurements is *not* the case [Verh. d. deutsch. phys. Ges. 13, 491 (1911)]. Compare also in this connexion the report of *E. Schrödinger* quoted in note 92.

100 *A. Einstein*, Ann. d. Phys. 22, 180, 800 (1907).

101 Cf. *A. Einstein*, Ann. d. Phys. 35, 683 ff. (1911), also the report by *Einstein* at the Solvay Congress in Brussels, 1911; see *A. Eucken*, Die Theorie der Strahlung und der Quanten. Abhandl. d. deutsch. Bunsen-Ges., Nr. 7 (W. Knapp, Halle 1914), pp. 330 *et seq.*

102 *A. Einstein*, Ann. d. Phys. 34, 170, 590 (1911); 35, 679 (1911).

103 The nature of the dependence of the frequency ν on the three

quantities A, ρ, κ may, according to *Einstein* (*loc. cit.*), be obtained by a simple dimensional calculation. If we assume that ν depends only on the mass m of the atoms, their distance apart d, and the compressibility κ of the body, then an equation of the following form must hold

$$\nu = C . m^x . d^y . \kappa^z.$$

C is here a numerical constant; x, y and z are numbers which remain to be determined.

The dimensions of the frequency $[\nu]$ are $[t^{-1}]$; the dimensions of m and d are $[m]$ and $[l]$, and the dimensions of the compressibility κ follow from its definition:

$$\kappa = \frac{\text{increase in volume}}{\text{increase in pressure} \times \text{original volume}}$$

κ has therefore the dimensions

$$\left[\frac{1}{\text{pressure}}\right] = \left[\frac{\text{surface}}{\text{force}}\right] = [lm^{-1}t^2].$$

We thus get the following dimensional equation

$$t^{-1} = C[m^x l^y l^z m^{-z} t^{2z}] = [m^{x-z} l^{y+z} t^{2z}].$$

Hence

$$x - z = 0; \qquad y + z = 0; \qquad 2z = -1$$

from which we get

$$x = -\tfrac{1}{2}; \qquad y = +\tfrac{1}{2}; \qquad z = -\tfrac{1}{2}$$

We have therefore,

$$\nu = Cm^{-\frac{1}{2}} \cdot d^{\frac{1}{2}} \kappa^{-\frac{1}{2}}.$$

Let N be *Avogadro's* number, i.e. the number of atoms in the gramme-atom. Then the atomic weight of the body is numerically equal to the mass of the gramme-atom, i.e.

$$A = mN.$$

If we imagine the atoms arranged upon a cubical space-lattice with sides d, then the density must satisfy the equation

$$\rho = \frac{m}{d^3}$$

from this it follows that

$$d = m^{\frac{1}{3}} \rho^{-\frac{1}{3}}$$

and hence

$$\frac{d}{m} = m^{-\frac{2}{3}} \rho^{-\frac{1}{3}} = A^{-\frac{2}{3}} N^{\frac{2}{3}} \rho^{-\frac{1}{3}}$$

from which, it follows that

$$\nu = CN^{\frac{1}{3}} \cdot A^{-\frac{1}{3}} \rho^{-\frac{1}{6}} \kappa^{-\frac{1}{2}} = \frac{C'}{A^{\frac{1}{3}} \rho^{\frac{1}{6}} \kappa^{\frac{1}{2}}}.$$

Einstein determines the factor C by assuming simply that only the twenty-six neighbouring atoms act upon the displaced atom.

104 *F. A. Lindemann*, Phys. Zeitschr. 11, 609 (1910). *Lindemann's* formula may be shortly deduced thus: Let $r = a \sin (2\pi\nu t)$ be the elongation of an atom which is vibrating with the amplitude a and the frequency ν. The mean energy of this atom is

$$E = \frac{m}{2}\left(\frac{dr}{dt}\right)^2 + \frac{m}{2} \cdot (2\pi\nu)^2 r^2 = \frac{m}{2} \cdot (2\pi\nu)^2 a^2 = 2\pi^2\nu^2 ma^2.$$

At the melting-point, according to *Lindemann's* conception, a is of the same order as d (distance apart of atoms). On the other hand, the mean energy of the atoms at high temperatures $= 3kT$, or, at the melting-point $3kT_s$. (The melting-point, as a rule, is high.) From this it follows that

$$2\pi^2\nu^2 md^2 = 3kT_s$$

$$\nu = \sqrt{\frac{3kT_s}{2\pi^2 md^2}} = \text{const. } T_s^{\frac{1}{2}} m^{-\frac{1}{2}} d^{-1}.$$

But we have (see note 103)

$$m = \frac{A}{N}; \qquad d = m^{\frac{1}{3}}\rho^{-\frac{1}{3}} = A^{\frac{1}{3}} N^{-\frac{1}{3}}\rho^{-\frac{1}{3}}.$$

Hence

$$\nu = \text{const. } T_s^{\frac{1}{2}} \cdot A^{-\frac{1}{2}} N^{\frac{1}{2}} A^{-\frac{1}{3}} N^{\frac{1}{3}}\rho^{\frac{1}{3}} = \text{const. } T_s^{\frac{1}{2}} \cdot A^{-\frac{5}{6}} \cdot \rho^{\frac{1}{3}}.$$

105 *E. Grüneisen*, Ann. d. Phys. 39, 291 *et seq.* (1912).

106 *E. Madelung*, Nachr. d. kgl. Ges. d. Wiss. zu Göttingen, mathem.-physikal. Klasse 1909, p. 100, and 1910, p. 1.

107 *W. Sutherland*, Phil. Mag. (6), 20, 657 (1910).

108 If n and κ are the coefficients of refraction and extinction of a substance respectively, then, according to *Maxwell's* Theory, its reflecting power is

$$\mathbf{R} = \frac{(n - 1)^2 + \kappa^2}{(n + 1)^2 + \kappa^2}.$$

If we require the point of maximum reflection, we have to form the equation $\dfrac{\partial \mathbf{R}}{\partial \nu} = 0$, which gives after reduction the following relation:

$$(n^2 - \kappa^2 - 1)\frac{\partial n}{\partial \nu} + 2n\kappa \frac{\partial \kappa}{\partial \nu} = 0.$$

From this we see that the position of maximum reflection does *not* coincide *exactly* with the position of maximum absorption $\left(\dfrac{\partial \kappa}{\partial \nu} = 0\right)$, but that it lies the nearer to it, the less the coefficient of refraction varies with the frequency. On the other hand, the point of maximum absorption lies, according to the dispersion theory, in the immediate neighbourhood of the natural frequency ν_r.

109 *H. Rubens* and *E. F. Nichols*, Wiedem. Ann. 60, 418 (1897). Also

H. Rubens and *H. Hollnagel*, Ber. d. kgl. preuss. Akad. d. Wiss. 1910, p. 45; *H. Hollnagel*, Dissert. Berlin 1910; *H. Rubens*, Ber. d. kgl. preuss. Akad. d. Wiss. 1913, p. 513; *H. Rubens* and *H. v. Wartenberg*, *ibid.*, 1914, p. 169.

As an example we give here the following small table in which λ denotes the wave-length of the " residual " rays, as given by the above investigators.

	λ		λ
NaCl	52 μ	TlCl	91·6μ
KCl	63·4μ	KBr	82·6μ
AgCl	81·5μ	AgBr	112·7μ
HgCl	98·8μ	TlBr	117 μ

110 Cf., however, note 108.

111 *W. Nernst* and *F. A. Lindemann*, Sitzungsber. d. kgl. preuss. Akad. d. Wiss. 1911, p. 494; *W. Nernst*, Ann. d. Phys. **36**, 426 (1911).

112 The following short table gives the values for ν which are calculated from *Einstein's* formula (35), *Lindemann's* formula (36), from the "residual rays" (see note 109), and from the observed atomic heat according to an empirical formula (40) proposed by *Nernst* and *Lindemann*. For more detailed data with, in part, corrected numerical factors see *C. E. Blom*, Ann. d. Phys. **42**, 1397 (1913).

Substance	ν_E	ν_L	ν residual rays	ν atomic heat (Nernst-Lindemann)
Al	$6·7 . 10^{12}$	$7·6 . 10^{12}$		$8·3 . 10^{12}$
Cu	$5·7 . 10^{12}$	$6·8 . 10^{12}$		$6·7 . 10^{12}$
Zn		$4·4 . 10^{12}$		$4·8 . 10^{12}$
Ag	$4·1 . 10^{12}$	$4·4 . 10^{12}$		$4·5 . 10^{12}$
Pb	$2·2 . 10^{12}$	$1·8 . 10^{12}$		$1·5 . 10^{12}$
Diamond		$32·5 . 10^{12}$		$40 . 10^{12}$
NaCl		$7·2 . 10^{12}$	$5·8 . 10^{12}$	$5·9 . 10^{12}$
KCl		$5·6 . 10^{12}$	$4·7 . 10^{12}$	$4·5 . 10^{12}$

113 *W. Nernst, F. Koref, F. A. Lindemann*, Untersuchungen über die spezifische Wärme bei tiefen Temperaturen. I. u. II. Sitzungsber. d. kgl. preuss. Akad. d. Wiss. 1910, 3 March.—*W. Nernst, idem* III., *ibid.*, 1911, 9 March.—*F. A. Lindemann, idem* IV., *ibid.*, 1911, 9 March.—*W. Nernst* and *F. A. Lindemann, idem* V., *ibid.*, 1911, 27 April.—*W. Nernst* and *F. A. Lindemann, idem* VI., *ibid.*, 1912, 12 Dec.—*W. Nernst, idem* VII., *ibid.*, 1912, 12 Dec.—*W. Nernst* and *F. Schwers, idem* VIII., *loc. cit.*, 1914.—*W. Nernst*, Der Energieinhalt fester Stoffe. Ann. d. Phys. **36**, 395 (1911).

114 *W. Nernst*, Die theoretischen und experimentellen Grundlagen des neuen Wärmesatzes. (W. Knapp, Halle 1918.)

115 The First Law states; If $d'Q$ is the heat *supplied* to a system, $d'A$

the work done *on* the system from outside, then the increase of energy U of the system is given by

$$dU = d'Q + d'A.$$

The Second Law states : if $d'Q$ is supplied reversibly at the temperature T, then $\frac{d'Q}{T}$ is the complete differential of the entropy S, hence

$$dS = \frac{d'Q}{T}.$$

Let us follow *Helmholtz* and introduce the " free energy " F defined by

$$F = U - T \cdot S.$$

Then it follows that

$$dF = dU - T \cdot dS - S \cdot dT = d'Q + d'A - T \cdot dS - S \cdot dT$$

i.e.

$$dF = d'A - S \cdot dT$$

for every reversible process.

If the process is *isothermal* ($dT = 0$) then it follows that $dF = d'A$ or, for a finite change of state, $F_2 - F_1 = A$. If we set $A' = -A$, so that A' is the work *gained*, we get

$$F_1 - F_2 = A'.$$

That is, the work gained in the *isothermal reversible* process—which is, as may be shown, the *maximum* obtainable—is equal to the decrease of free energy.

Further, it follows, since at constant volume V the work $d'A = 0$, that

$$\left(\frac{\partial F}{\partial T}\right)_V = -S \quad \text{or} \quad T \cdot \left(\frac{\partial F}{\partial T}\right)_V = -TS = F - U.$$

Therefore, formulating these expressions for two states, we get

$$T\left[\frac{\partial(F_1 - F_2)}{\partial T}\right]_V = (F_1 - F_2) - (U_1 - U_2)$$

or, finally, if we write for short $U_1 - U_2 = U'$

$$A' - U' = T\left(\frac{\partial A'}{\partial T}\right)_V$$

an equation much used in physical chemistry.

Since, now, according to *Nernst's* heat theorem,

$$\left(\frac{\partial A'}{\partial T}\right)_{\lim T=0} = 0$$

$(A' - U')$ vanishes for $T = 0$, being above the first order.

Hence

$$\lim_{T=0} \frac{\partial(A' - U')}{\partial T} = 0$$

and hence also

$$\lim_{T=0} \frac{\partial U'}{\partial T} = 0 \quad \text{or} \quad \lim_{T=0} \frac{\partial U_1}{\partial T} = \lim_{T=0} \frac{\partial U_2}{\partial T}.$$

This is equation (89) of the text.

From $\frac{\partial F}{\partial T} = - S$, it follows further that $\frac{\partial(F_1 - F_2)}{\partial T} = S_2 - S_1$, or

$$S_2 - S_1 = \frac{\partial A'}{\partial T}$$

and hence *Nernst's* Theorem may be formulated thus

$$\lim_{T=0} (S_2 - S_1) = 0$$

that is, *in the neighbourhood of the absolute zero all processes proceed without change of entropy.*

116 Cf., for example, *M. Planck*, Lectures on Thermodynamics. *Planck* goes further than *Nernst* inasmuch as he postulates that not only the difference of the entropies $S_2 - S_1$ is zero at absolute zero (see previous note) but also that the individual values themselves become zero. Hence, according to *Planck*, at the absolute zero of temperature *the entropy of every chemically homogeneous body is equal to zero.* From this the conclusion given in the text,

$$\lim_{T=0}\left(\frac{\partial U}{\partial T}\right) = 0$$

may be deduced immediately. It follows from the relation (occurring in the last note)

$$F - U = - TS$$

and from *Planck's* version of *Nernst's* Theorem, that $F - U$ vanishes for $T = 0$, being of higher order than the first.
Hence

$$\lim_{T=0}\frac{\partial(F - U)}{\partial T} = 0 \quad \text{or} \quad \lim_{T=0}\left(\frac{\partial U}{\partial T} + S\right) = 0$$

or, finally,

$$\lim_{T=0}\left(\frac{\partial U}{\partial T}\right) = 0.$$

117 For low temperatures, that is, for high values of $x = \frac{h\nu}{kT}$, *Einstein's* formula (34) takes the following form: $C_v = 3Rx^2e^{-x}$. The falling-off at low temperatures therefore follows an exponential law; more exactly, it varies as

$$\frac{1}{T^2}e^{-\frac{\text{const}}{T}}.$$

118 *W. Nernst* and *F. A. Lindemann*, Sitzungsber. d. kgl. preuss. Akad. d. Wiss. 1911, p. 494; Zeitschr. f. Elektrochemie, **17**, 817 (1911).

119 *A. Einstein*, Ann. d. Phys. **35**, 679 (1911).

120 For if we regard the atoms as mass-points, then each atom has three degrees of freedom; the whole body has therefore $3N$ degrees of freedom. As is proved in mechanics, however (cf. *R. H. Weber* and *R. Gans*,

Repertorium der Physik Bd. I. pp. 175 *et seq.*), a mechanical system of $3N$ degrees of freedom has $3N$ natural frequencies, and *the most general small motion of each atom consists in a superposition of these $3N$ natural frequencies.*

121 *P. Debye*, Ann. d. Phys. 39, 789 (1912).

122 *M. Born* and *Th. v. Kármán*, Phys. Zeitschr. 13, 297 (1912); 14, 15, 65 (1913). Cf. also *M. Born*, Ann. d. Phys. 44, 605 (1914); *M. Born*, Dynamik der Kristallgitter (Teubner, Leipzig and Berlin 1915).

123 Cf., for example, *R. Ortvay*, Über die Abzählung der Eigenschwingungen fester Körper. Ann. d. Phys. 42, 745 (1913).

Ortvay considers the natural frequencies of an elastic cube, each side of which has the length L. There are found to be three groups of natural frequencies. The first two groups are the *transversal* frequencies, the third group is the group of the longitudinal frequencies. That the transversal frequencies form two groups (moreover identical) is easily seen. For in the case of a transversal vibration, which is propagated in, say, the direction of the x-axis, *two* equal alternatives are probable, namely, that the particles vibrate parallel to the y- or to the z-axis. In the case of the longitudinal oscillations, however, there is naturally only *one* group; for in the case of propagation along the x-axis there is only *one* possibility, namely, that the particles vibrate parallel to the x-axis. The frequencies of the first two groups are characterised by the values

$$\nu_1 = \nu_2 = c_t \frac{\sqrt{\mathbf{a}^2 + \mathbf{b}^2 + \mathbf{c}^2}}{2L}$$

the third group by

$$\nu_3 = c_l \frac{\sqrt{\mathbf{a}^2 + \mathbf{b}^2 + \mathbf{c}^2}}{2L}.$$

Here c_t and c_l are the velocities of propagation of transversal and longitudinal waves in the body, whereas $\mathbf{a}$, $\mathbf{b}$, $\mathbf{c}$ are arbitrary positive whole numbers. If therefore we give $\mathbf{a}$, $\mathbf{b}$, $\mathbf{c}$ all possible values in all possible combinations, we get all the possible transversal and longitudinal natural frequencies, which together form the elastic spectrum of the cube. If now we inquire how many transversal natural frequencies of the first group fall below ν, this means nothing else than inquiring how many trios of values ($\mathbf{a}$, $\mathbf{b}$, $\mathbf{c}$) fulfil the condition

$$c_t \frac{\sqrt{\mathbf{a}^2 + \mathbf{b}^2 + \mathbf{c}^2}}{2L} < \nu$$

i.e.

$$\sqrt{\mathbf{a}^2 + \mathbf{b}^2 + \mathbf{c}^2} < \frac{2L\nu}{c_t}.$$

Imagine $\mathbf{a}$, $\mathbf{b}$, $\mathbf{c}$ as co-ordinates of a point in space. Then all possible trios ($\mathbf{a}$, $\mathbf{b}$, $\mathbf{c}$) of values are represented by the total "lattice-points" of the positive space octant, and the above question is answered by counting how many lattice-points are at a distance less than $\frac{2L\nu}{c_t}$ from the origin $(0, 0, 0)$.

All these lattice-points lie within the positive octant of the sphere whose radius is $\frac{2L\nu}{c_t}$. Since now *one* lattice-point is assigned to every volume of magnitude 1—namely, every elementary cube—the required number of lattice-points, provided that it is sufficiently large, is equal to the volume of the positive spherical octant of radius $\frac{2L\nu}{c_t}$, i.e. is equal to

$$\frac{1}{8}\cdot\frac{4\pi}{3}\left(\frac{2L\nu}{c_t}\right)^3 = \frac{4\pi}{3}\frac{L^3\nu^3}{c_t^3}.$$

If $V = L^3$ is the volume of the given cubical body, then the number of the transverse natural frequencies below ν belonging to the first group is

$$Z_1 = \frac{4\pi}{3}V\frac{\nu^3}{c_t^3}.$$

The number belonging to the second group is the same, that is

$$Z_2 = Z_1 = \frac{4\pi}{3}V\frac{\nu^3}{c_t^3}.$$

Finally, the number of the longitudinal frequencies corresponding to these is

$$Z_3 = \frac{4\pi}{3}V\frac{\nu^3}{c_l^3}.$$

We thus get for the total of all natural frequencies below ν

$$Z = Z_1 + Z_2 + Z_3 = \frac{4\pi}{3}V\left(\frac{2}{c_t^3}+\frac{1}{c_l^3}\right)\nu^3.$$

The total of natural frequencies in the interval $\nu \ldots \nu + d\nu$ follows by differentiation with respect to ν

$$Z(\nu)d\nu = 4\pi V\left(\frac{2}{c_t^3}+\frac{1}{c_l^3}\right)\nu^2 d\nu$$

and this is just formula (43) of the text.

124 In formula (43) for $Z(\nu)d\nu$ let us replace, according to formula (44) of the text, the factor

$$4\pi V\left(\frac{1}{c_l^3}+\frac{2}{c_t^3}\right) \text{ by } \frac{9N}{\nu_m^3}.$$

Then it follows that

$$C_v = \frac{9R}{\nu_m^3}\int_0^{\nu_m}\frac{\left(\frac{h\nu}{kT}\right)^2 e^{\frac{h\nu}{kT}}}{\left(e^{\frac{h\nu}{kT}}-1\right)}\nu^2 d\nu.$$

If we now set $\frac{h\nu}{kT} = \kappa$ and $\frac{h\nu_m}{kT} = \kappa_m$, we get

$$C_v = \frac{9R}{\kappa_m^3} \int_0^{x_m} \frac{x^4 e^x dx}{(e^x - 1)^2}.$$

125 A table showing how the *Debye* function C_v depends on x_m is given by *Nernst* (Die theoretischen und experimentellen Grundlagen des neuen Wärmesatzes. W. Knapp, Halle 1918, p. 201). In it the simple *Einstein* function [formula (34) of the text] is also tabulated.

126 If T is great, then x_m is small compared with 1; then we may replace in the integral of (45) e^x by 1 in the numerator, and $e^x - 1$ by x in the denominator. It then follows that

$$C_v = \frac{9R}{x_m^3} \int_0^{x_m} \frac{x^4 dx}{x^2} = \frac{9R}{x_m^3} \cdot \frac{x_m^3}{3} = 3R.$$

127 If T is small, then x_m is large, and we may replace the upper limit of the integral as a first approximation by ∞. The integral will thus become a numerical constant independent of x_m, and it follows that

$$C_v = \frac{9R}{x_m^3} \cdot \text{const.} = \frac{9R \cdot \text{const.} \cdot \nu_m^3}{h^3 k^3} \cdot T^3 = \text{const.} \cdot T^3.$$

128 From the theory of elasticity it follows that

$$c_l = \sqrt{\frac{3(1-\sigma)}{(1+\sigma)\kappa\rho}} \qquad \text{and} \qquad c_t = \sqrt{\frac{3(1-2\sigma)}{2(1+\sigma)\kappa\rho}}$$

where κ is the compressibility, ρ the density, and σ the ratio

$$\frac{\text{transverse contraction}}{\text{longitudinal dilatation}}.$$

If we insert these values in (44) and note further that $V = \frac{A}{\rho}$, formula (46) of the text follows.

129 As the number of frequencies below ν is proportional to ν^3, we get, for example, the following picture: if we divide the interval from 0 to ν_m into 10 parts, and if only *one* natural frequency lies in the first division, then in the following divisions there will be 7, 19, 37, 61, 91, 127, 169, 217, 271 natural frequencies; i.e. the natural frequencies crowd continually closer together.

130 P. *Debye*, Ann. d. Phys. **39**, 789 (1912); W. *Nernst* and F. A. *Lindemann*, Sitzungsber. d. Berl. Akad. d. Wiss. 1912, p. 1160.

131 A. *Eucken* and F. *Schwers*, Verhandl. d. deutsch. physikal. Ges. **15**, 578 (1913); W. *Nernst* and F. *Schwers*, Sitzungsber. d. Berl. Akad. d. Wiss. 1914, p. 355; P. *Grünther*, Ann. d. Phys. **51**, 828 (1916); W. H.

Keesom and *Kamerlingh-Onnes*, Amsterdam Proc. **17**, 894 (1915). Cf. also the graphic tables by *E. Schrödinger*, Phys. Zeitschr. **20**, 498 (1919).

132 If we introduce into equation (44) of the text,

$$\frac{4\pi V}{3}\left(\frac{1}{c_l^3} + \frac{2}{c_t^3}\right)\nu_m^3 = 3N$$

a " mean acoustic velocity " $\bar{c}$, by the obvious definition

$$\frac{3}{\bar{c}^3} = \frac{1}{c_l^3} + \frac{2}{c_t^3}$$

then for the order of magnitude of the smallest wave-length λ_{min}, there follows

$$\frac{\bar{c}}{\nu_m} \backsim \lambda_{min} \backsim \sqrt{\frac{4\pi V}{3N}}.$$

If now the atoms in the cubical space-lattice, for example, are arranged so as to be a distance a apart, then $Na^3 = V$, and hence

$$\lambda_{min} \backsim \sqrt{\frac{4\pi}{3}} \cdot a.$$

133 For references see note 122.

134 Cf. *Born*, Dynamik der Kristallgitter, § 19.

135 *F. Haber*, Verh. d. deutsch. phys. Ges. **13**, 1117 (1911).

For if the atomic residue (mass m) and the electron (mass μ) are held to their zero positions by forces of the same order of magnitude, and if they vibrate independently of one another (a simplifying supposition) the equation of vibration of the atom is $m\ddot{x} + a^2 x = 0$, the solution of which is

$$x = A \sin\left(\frac{a}{\sqrt{m}}t\right).$$ The infra-red frequency of the atom is, therefore,

$\nu_r = \dfrac{a}{2\pi\sqrt{m}}$, and correspondingly, the ultra-violet frequency of the electron

is $\nu_e = \dfrac{a}{2\pi\sqrt{\mu}}$. Hence *Haber's* Law follows: $\nu_r : \nu_e = \sqrt{\mu} : \sqrt{m}$. The general space-lattice theory of M. *Born* confirms this law and shows that in the lattice, too, atomic residues and electrons appear upon an equal footing, and are acted upon by forces of *the same order of magnitude*.

136 Cf. *M. Born* and *Thos. v. Kármán*, Phys. Zeitschr. **13**, 297 (1912).

We may treat this problem, which is of course one-dimensional, most simply thus: If we imagine an endless chain of points of equal mass m disposed along the x-axis at a distance apart a, and if we suppose for simplicity that each mass-point only acts upon its two neighbours, then the equation of motion of the nth point is

$$m\ddot{x}_n = a(x_{n+1} - x_n) - a(x_n - x_{n-1}) = a(x_{n+1} + x_{n-1} - 2x_n).$$

Here a is a constant, and n can assume all values between $+\infty$ and $-\infty$.

As a solution let us set for trial

$$x_n = A \sin\left(2\pi\nu t - n\frac{2\pi a}{\lambda}\right).$$

This represents a process which is periodic in space and time, that is, a wave which is propagated along the chain in the direction of increasing x. The frequency of this wave is ν, its length is λ. Then, if after p points the same displacement is to recur, pa must $= \lambda$, and hence it actually follows that

$$x_{n+p} = A \sin\left[2\pi\nu t - (n+p)\frac{2\pi a}{\lambda}\right]$$

$$= A \sin\left(2\pi\nu t - n\frac{2\pi a}{\lambda}\right) = x_n.$$

In order to find the relation between ν and λ (that is, the "law of dispersion"), let us insert the above formula in the equation of motion. Then it follows that

$$- m(2\pi\nu)^2 A \sin\left(2\pi\nu t - n\frac{2\pi a}{\lambda}\right)$$

$$= aA\left\{\sin\left[2\pi\nu t - (n+1)\frac{2\pi a}{\lambda}\right] + \sin\left[2\pi\nu t - (n-1)\frac{2\pi a}{\lambda}\right]\right.$$

$$\left. - 2\sin\left[2\pi\nu t - n\frac{2\pi a}{\lambda}\right]\right\}$$

$$= - 2aA \sin\left[2\pi\nu t - n\frac{2\pi a}{\lambda}\right] \cdot \left(1 - \cos\frac{2\pi a}{\lambda}\right).$$

That is,

$$(2\pi\nu)^2 m = 4a \sin^2\left(\frac{\pi a}{\lambda}\right).$$

$$\nu = \frac{1}{\pi}\sqrt{\frac{a}{m}} \sin\left(\frac{\pi a}{\lambda}\right) = \nu_m \sin\left(\frac{\pi a}{\lambda}\right), \text{ if we set } \frac{1}{\pi}\sqrt{\frac{a}{m}} = \nu_m.$$

137 Cf. *Born*, Dynamik der Kristallgitter, p. 51.

From the special case treated in the previous note, we also recognise the truth of law (49); for if λ is much greater than a, the dispersion law takes the form $\nu = \nu_m \frac{\pi a}{\lambda} = \frac{q}{\lambda}$, where $q = \nu_m \pi a$, represents the velocity of propagation of the wave, and this is independent of the wave-length.

138 The statement that a given direction lies in the element of solid angle $d\Omega$ is intended to convey the following sense: about an arbitrary origin O describe a "unit sphere," i.e. a sphere of radius 1. Now let a cone of infinitely small angle be constructed of rays passing through O, the point of the cone lying at O. Let this cone cut out of the surface of the unit sphere a small element of surface $d\Omega$. Now let the parallel ray to the "given direction" be drawn through O (here, for example, the wave-normals). If this ray lies in the cone just constructed, then we say that the "given direction" lies in the elementary solid angle $d\Omega$.

139 The capacity for heat of a certain finite body is that amount of heat which must be imparted to *the whole body* in order that its temperature be raised by 1° C. If M is the mass of the body, and c its specific heat, then its capacity for heat is

$$\Gamma = cM.$$

From the mean energy content $\bar{E}$ of *the whole body*, Γ follows by differentiation with respect to the temperature

$$\Gamma = \frac{d\bar{E}}{dT}.$$

140 This somewhat complicated calculation runs as follows: we start from the formula

$$\Gamma_v^{(1)} = kV \sum_{i=1}^{3} \int_0^{4\pi} d\Omega \int_{\lambda_m(n)}^{\infty} \frac{d\lambda}{\lambda^4} \cdot \frac{\left(\frac{h\nu_i}{kT}\right)^2 e^{\frac{h\nu_i}{kT}}}{\left(e^{\frac{h\nu_i}{kT}} - 1\right)^2}$$

and first replace λ by $\frac{q_i(n)}{\nu_n}$. Thus we get

$$\frac{d\lambda}{\lambda^4} = -\frac{\nu_i^2 d\nu_i}{q_i^3(n)}$$

and the integral with respect to λ is transformed into one with respect to ν_i. The limits of this integral are

$$\nu_i = \frac{q_i(n)}{\lambda_m(n)} \qquad [\text{corresponding to } \lambda = \lambda_m(n)]$$

and

$$\nu_i = 0 \qquad (\text{corresponding to } \lambda = \infty).$$

If we further set

$$\frac{h\nu_i}{kT} = x \quad \text{and} \quad \frac{hq_i(n)}{kT\lambda_m(n)} = x_i(n)$$

we get

$$\Gamma_v^{(1)} = kV \frac{k^3 T^3}{h^3} \sum_{i=1}^{3} \int_0^{4\pi} \frac{d\Omega}{q_i^3(n)} \int_0^{x_i(n)} \frac{x^4 e^x dx}{(e^x - 1)^2}.$$

In place of the quantities $q_i(n)$ and $\lambda_m(n)$ which still depend essentially on the direction, let certain *mean values* be introduced. Firstly, let us set

$$\frac{1}{\bar{q}_i^3} = \frac{1}{4\pi} \int_0^{4\pi} \frac{d\Omega}{q_i^3(n)} \qquad (i = 1, 2, 3 \ldots).$$

In this way three mean acoustic velocities $\bar{q}_1, \bar{q}_2, \bar{q}_3$, independent of the direction, are defined. We further introduce in place of $\lambda_m(n)$ a mean value independent of the direction, in the following manner. In deducing formula (55) we saw that

$$V \int_0^{4\pi} d\Omega \int_{\lambda_m(n)}^{\infty} \frac{d\lambda}{\lambda^4} = N.$$

If we carry out the integration with respect to λ, we get

$$\frac{V}{3}\int_0^{4\pi}\frac{d\Omega}{\lambda_m^3(n)} = N$$

i.e.

$$\int_0^{4\pi}\frac{d\Omega}{\lambda_m^3(n)} = \frac{3N}{V}.$$

Now, in a way analogous to that used for the acoustic velocities q, we set

$$\frac{1}{\overline{\lambda_m}^3} = \frac{1}{4\pi}\int_0^{4\pi}\frac{d\Omega}{\lambda_m^3(n)} = \frac{3N}{4\pi V}$$

Hence

$$\overline{\lambda_m}^3 = \sqrt{\frac{4\pi V}{3N}}.$$

Into $x_i(n) = \frac{hq_i(n)}{kT\lambda_m(n)}$ we introduce in place of $q_i(n)$ and $\lambda_m(n)$ the mean values q_i and λ_m, which are independent of direction; thereby $x_i(n)$ also becomes *independent of direction*, and is transformed into

$$x_i = \frac{h\overline{q_i}}{kT\overline{\lambda_m}} = \frac{h\overline{q_i}}{kT}\sqrt[3]{\frac{3N}{4\pi V}}.$$

It follows that

$$\Gamma_v^{(1)} = 4\pi k V\frac{k^2 T^3}{h^3}\sum_{i=1}^3\frac{1}{\overline{q_i}^3}\int_0^{\overline{x_i}}\frac{x^4 e^x dx}{(e^x-1)^2}$$

$$= 3Nk\sum_{i=1}^3\frac{1}{\overline{x_i}^3}\cdot\int_0^{\overline{x_i}}\frac{x^4 e^x dx}{(e^x-1)^2}$$

$$= 3R\cdot\sum_{i=1}^3\frac{1}{\overline{x_i}^3}\cdot\int_0^{\overline{x_i}}\frac{x^4 e^x dx}{(e^x-1)^2}.$$

141 At the lowest temperatures

$$\overline{C}_v = \frac{1}{3s}\cdot\sum_1^3 D(\overline{x_i}) = \frac{1}{3s}\cdot 9R\sum_{i=1}^3\frac{1}{\overline{x_i}^3}\int_0^\infty\frac{x^4 e^x dx}{(e^x-1)^2}$$

Now the value of the integrals $= \frac{4}{15}\pi^4$. If we further set $R = Nk$, and for $\overline{x_i}$ the value (59), we get

$$\overline{C}_v = \frac{16\pi^3 k^4 V T^3}{15h^3 s} \cdot \sum_1^3 \frac{1}{\overline{q_i^3}} \cdot$$

If we introduce in place of the three acoustic velocities $\overline{q_1}, \overline{q_2}, \overline{q_3}$ a mean acoustic velocity $\overline{q}$ by means of the definition

$$\frac{3}{\overline{q}^3} = \sum_{i=1}^3 \frac{1}{\overline{q_i^3}} = \sum_{i=1}^3 \frac{1}{4\pi} \int_0^{4\pi} \frac{d\Omega}{q_i^3(n)} \quad \text{(see note 140)}$$

it follows that

$$\overline{C}_v = \frac{16\pi^3 k^4 V T^3}{5h^3 s \overline{q}^3}.$$

Finally for $\frac{V}{s}$ we can write $\overline{V}_A$ (mean atomic volume) and thus get the formula

$$\overline{C}_v = \frac{16\pi^3 k^4 \overline{V}_A T^3}{5h^3 \overline{q}^3}.$$

[142] H. Thirring, Phys. Zeitschr. 14, 867 (1913) ; 15, 127, 180 (1914).

[143] M. Born and Th. v. Kármán, Phys. Zeitschr. 14, 15 (1913).

[144] Cf. note 132.

[145] Cf. note 128.

[146] A. Eucken, Verhandl. d. deutsch. physikal. Ges. 15, 571 (1913). Cf. also A. Eucken, Die Theorie der Strahlung und der Quanten (W. Knapp, Halle 1914), pp. 386 et seq., Appendix.

[147] Cf. A. Eucken, Die Theorie der Strahlung und der Quanten (W. Knapp, Halle 1914), p. 387.

[148] To calculate the mean acoustic velocity $\overline{q}$, the relation given in note 141 is used

$$\frac{3}{\overline{q}^3} = \sum_{i=1}^3 \frac{1}{4\pi} \int_0^{4\pi} \frac{d\Omega}{q_i^3(n)}$$

We have therefore to obtain from the "dispersion equation" of the crystal in question (for long waves) the values of the three acoustic velocities $q_1(n)$, $q_2(n)$, $q_3(n)$ as functions of the wave-direction; $\overline{q}$ is then obtained from the above formula by integration over all directions and finally summation.

[149] L. Hopf and G. Lechner, Verhandl. d. deutsch. physikal. Ges. 16, 643 (1914).

[150] The following short table is taken from the paper of Hopf and Lechner cited in note 149 :

Crystal	$\bar{q}$ calc. from $\bar{C}_v$	$\bar{q}$ calc. from elastic data
Sylvin . . .	$2{\cdot}36 \, . \, 10^5$	$2{\cdot}03 \, . \, 10^5$
Rock salt . .	$2{\cdot}82 \, . \, 10^5$	$2{\cdot}72 \, . \, 10^5$
Fluor-spar . .	$4{\cdot}02 \, . \, 10^5$	$3{\cdot}82 \, . \, 10^9$
Pyrites . .	$5{\cdot}43 \, . \, 10^5$	$5{\cdot}12 \, . \, 10^5$

151 *W. Nernst*, Vorträge über die kinetische Theorie der Materia und der Elektrizität. Wolfskehl-Kongress 1913 in Göttingen (Teubner, Leipzig and Berlin 1914), pp. 63 *et seq.*

152 *W. Nernst, ibid.*, pp. 81 *et seq.*

153 *E. Schrödinger*, Phys. Zeitschr. 20, 503 (1919). *Schrödinger* correctly points out that—apart from the substitution of *one single* mean $\bar{x}$ for the three quantities x_i in the *Debye* terms—the approximation above all in the second part of C_v (i.e. the replacement of the $3(s-1)$ frequencies $\nu_4 \ldots \nu_3$ by the *constants* $\nu_4^0 \ldots \nu_{3s}^0$) may not be permissible in many cases: namely, in those cases in which the masses of the various kinds of atoms are not very different from one another. If we were to allow —so he argues—the masses of the different kinds of atoms and the forces acting upon them gradually to become equal to one another, a simple atomic lattice would result, and during this process the $3(s-1)$ branches of the spectrum, which correspond to the second type of motion, would merge into the three first branches. "They cannot therefore even be approximately monochromatic if the masses differ only slightly."

154 *H. Thirring*, Phys. Zeitschr. 15, 127, 180 (1914).

155 *M. Born*, Ann. d. Phys. 44, 605 (1914).

156 *E. Grüneisen*, Ann. d. Phys. 39, 257 (1912).

157 *S. Ratnowski*, Verhandl. d. deutsch. physikal. Ges. 15, 75 (1913).

158 Vorträge über die kinetische Theorie der Materie und der Elektrizität. Wolfskehl-Kongresz zu Göttingen, 1913. (Teubner, Leipzig and Berlin 1914), Vortrag *P. Debye.*

159 If U is the energy, and S the entropy of the system, then the "free energy" is defined according to *Helmholtz* by the relation

$$F = U - S \, . \, T.$$

It then follows from note 115 that

$$dF = d'A - S \, . \, dT$$

where $d'A$ is the work done from without. If we set in the usual way

$$d'A = - pdV \qquad (p = \text{pressure}, \; V = \text{volume})$$

then

$$dF = - pdV - SdT.$$

From this we get immediately the equation (66) in the text

$$\left(\frac{\partial F}{\partial V}\right)_T = - p.$$

Similarly,

$$\left(\frac{\partial F}{\partial T}\right)_r = - S$$

and hence

$$U = F + T \cdot S = F - T\left(\frac{\partial F}{\partial T}\right)_r.$$

160 *P. Debye*, loc. cit., note 158.

161 *E. Grüneisen*, Ann. d. Phys. **26**, 211 (1908); **33**, 65 (1910); **39**, 285 (1912).

162 *P. Debye*, loc. cit., note 158.

163 *A. Eucken*, Ann. d. Phys. **34**, 185 (1911); Verhandl. d. deutsch. physikal. Ges. **13**, 829 (1911).

164 *P. Drude*, Ann. d. Phys. **1**, 566 (1900).

165 *E. Riecke*, Wiedem. Ann. **66**, 353, 545 (1898).

166 Cf., for example, *H. A. Lorentz*, The Theory of Electrons (Teubner, Leipzig, and Berlin 1909).

167 Let q be the average velocity of the electrons along the free path l. Then the electron takes the time $\tau = \dfrac{l}{q}$ to pass over this free path. During this time it is exposed to the electrical force $\mathbf{E}$ of the external field. Its increase in velocity due to this force is at the commencement of the free path $= 0$, at the end of it $= \dfrac{e\mathbf{E}\tau}{m}$, where e and m are the charge and mass of the electron respectively. In the mean, therefore, the small additional velocity generated by the field is $\Delta q = \dfrac{e\mathbf{E}\tau}{2m} = \dfrac{e\mathbf{E}l}{2mq}$. The electrons stream *unidirectionally* with this velocity against the field. If $\mathbf{N}$ is the number of electrons per unit volume, then through unit area of the surface there streams per second a quantity of electricity $N e \Delta q = \dfrac{N e^2 l \mathbf{E}}{2mq}$. This is, however, the "current density" $\mathbf{I}$ which is known to be connected with the field $\mathbf{E}$ by the relation $\mathbf{I} = \sigma\mathbf{E}$. The expression (67) for the conductivity σ therefore follows.

A more thorough treatment is due to *H. A. Lorentz* (see note 166). He does not give the electrons a single velocity q, but introduces *Maxwell's* supposition, known from the kinetic theory of gases, that all possible velocities occur, which are distributed among the electrons according to a fixed law, the so-called *Maxwell* Law of Distribution. He thus obtained a formula of the following form:

$$\sigma = 2\sqrt{\frac{2}{3\pi}}\,\frac{N e^2 l}{mq}$$

which therefore only differs by a numerical factor from *Drude's* formula (67); here $q = \sqrt{\overline{q^2}}$, the root mean square of the velocity.

168 Let a temperature gradient along the x axis be present in the piece of metal. Let a section be taken (see Fig. 12) at right angles to the

11

x axis; we shall calculate the energy transport across this section per second. If we suppose that $\frac{1}{3}$ of all electrons wander in each of the three directions in space, then $\frac{1}{6}$ move in the positive x direction; and further,

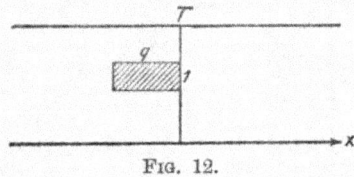

Fig. 12.

the number of electrons which pass through the unit of surface in one second, will be all those which are contained in the small shaded cylinder with the base surface area 1 and the height q (velocity), namely, $\frac{1}{6}Nq$. We also make the supposition, usual in the theory of gases (although not strictly true), that the energy, which each electron transports through the cross-section, has the value corresponding to that which it had at the point where it last collided.

Now the energy in the section itself at temperature T is equal to $\frac{3}{2}kT$, and hence the energy $= \frac{3}{2}kT \pm \dfrac{\partial(\frac{3}{2}kT)}{\partial x} . l$ at the points which lie at a distance l in front of and behind the section. Here, on the average, the electrons coming from the right and the left meet with their last collisions.

The energy transport per second through unit of cross-section is therefore

$$\left(\tfrac{3}{2}kT - \tfrac{3}{2}k\frac{\partial T}{\partial x} l \right)\tfrac{1}{6}Nq - \left(\tfrac{3}{2}kT + \tfrac{3}{2}k\frac{\partial T}{\partial x} l \right)\tfrac{1}{6}Nq$$

$$= -\tfrac{1}{2}Nlqk\frac{\partial T}{\partial x} = -\gamma\frac{\partial T}{\partial x}.$$

Hence $\gamma = \frac{1}{2}Nlqk$ is the coefficient of thermal conductivity.

Here also *H. A. Lorentz* has deepened the theory by taking the distribution of velocity into account, and finds that

$$\gamma = \tfrac{4}{3}\sqrt{\frac{2}{3\pi}}\,Nlqk$$

where again $q = \sqrt{\overline{q^2}}$

169 G. *Wiedemann* and R. *Franz*, Poggend. Ann. **89**, 497 (1853); L. *Lorenz*, Wiedem. Ann. **13**, 422, 582 (1881). Cf. also G. *Kirchhoff* and G. *Hansemann*, Wiedem. Ann. **13**, 417 (1881); W. *Jaeger* and H. *Diesselhorst*, Abh. d. phys. techn. Reichsanstalt **3**, 269 (1900).

The following short table is taken from the paper of the two investigators last named; it gives the ratio $\dfrac{\gamma}{\sigma}$ for various metals at a temperature of 18° C.

Metal	$\frac{Y}{\sigma} \cdot 10^{-10}$
Al	6·36
Cu	6·65
Ag	6·86
Au	7·09
Zn	6·72
Pb	7·15
Pt	7·53
Bi	9·64

170 It follows from (67) by setting $\frac{1}{2}mq^2 = \frac{3}{2}kT$, that is, $q = \sqrt{\frac{3kT}{m}}$, that

$$l = 2\sigma \frac{\sqrt{3kTm}}{Ne^2}.$$

Now, let N be the number of atoms per unit volume, N^* the number of atoms in a gramme-atom (*Avogadro's* number). If, further, A is the atomic weight, M the mass of an atom, and ρ the density, then

$$\begin{cases} A = MN^* \\ \rho = MN \end{cases}$$

therefore

$$N = N^* \frac{\rho}{A}.$$

We next assume that $\mathbf{N}$, the number of electrons per unit of volume, is small compared with N, say

$$\mathbf{N} = \frac{1}{100}N = \frac{1}{100} N^* \frac{\rho}{A}.$$

If we insert this value, then we get for the free path

$$l = \frac{200\,\sigma A \sqrt{3kTm}}{N^* \rho e^2}.$$

We shall make a rough calculation for copper at 0° C. We have

$$\sigma \backsim 5\cdot4 \cdot 10^{17} \text{ (in electrostatic units)}$$
$$A = 63\cdot57$$
$$k = 1\cdot4 \cdot 10^{-16}$$
$$T = 273$$
$$m = 0\cdot9 \cdot 10^{-27}$$
$$N^* = 6\cdot1 \cdot 10^{23}$$
$$\rho = 8\cdot9$$
$$e = 4\cdot77 \cdot 10^{-10}.$$

With these values we get

$$l \text{ is of the order } 5\cdot7 \cdot 10^{-5}.$$

Since the atomic distance is of the order of magnitude $2 \cdot 10^{-8}$, the electrons would therefore only suffer collision after passing many thousands of atoms. This is unacceptable, since the "radius of molecular action" of the atoms itself has dimensions which fall within the order of magnitude of about 10^{-8}.

171 *H. A. Lorentz, loc. cit.*, note 43.

172 *J. J. Thomson*, The Corpuscular Theory of Matter.

173 *H. Kammerlingh-Onnes*, Leiden Communicat. 1913, 133.

174 *C. H. Lees*, Phil. Trans. (A) 208, 381–443 (1908).

175 *W. Meissner*, Ann. d. Phys. 47, 1001 (1915).

176 *W. Nernst*, Berl. Ber. 1911, p. 310.

177 *H. Kammerlingh-Onnes*, Leiden Communicat. 119, 22 (1911).

178 *F. A. Lindemann*, Berl. Ber. 1911, p. 316.

179 *W. Wien*, Berl. Ber. 1913, p. 184. Cf. also Vorlesungen über neuere Probleme der theoretischen Physik. (Teubner, Leipzig and Berlin 1913.) 3. Vorlesung.

180 If s is the radius of atomic action, N the number of stationary atoms per unit of volume, then, according to a well-known result of the kinetic theory of gases, the mean free paths of the electrons

$$l = \frac{1}{\pi N s^2}.$$

Let us set

$$s = s_0 + a$$

where s_0 is the radius of atomic action for $T = 0$, that is, when the atoms are at rest; let a be the amplitude of atomic vibration. Now the mean energy E of this vibration (frequency ν), on the one hand, $= \frac{M}{2}(2\pi\nu)^2 a^2$ (M is the atomic mass); on the other hand, it is, according to *Planck-Einstein*,

$$= \frac{3h\nu}{e^{\frac{h\nu}{kT}} - 1}.$$

From this it follows that

$$a = \frac{1}{\pi}\sqrt{\frac{3h}{2M\nu\left(e^{\frac{h\nu}{kT}} - 1\right)}}.$$

Now, according to formula (67) of the text, the resistance

$$W = \frac{1}{\sigma} = \frac{2mq}{Ne^2 l}.$$

If we here set for q the value $\sqrt{\frac{3kT}{m}}$ (cf. note 170), and for N, according to *J. J. Thomson's* supposition, $a\sqrt{T}$, and for $\frac{1}{l}$ the value

$$\pi N s^2 = \pi N(a^2 + 2as_0 + s_0^2)$$

it follows that

$$W = \frac{2\pi N \sqrt{3km}}{e^2 \alpha} \left\{ \frac{\frac{3h}{2M\nu\pi^2}}{e^{\frac{h\nu}{kT}} - 1} + \frac{2s_0}{\pi} \frac{\sqrt{\frac{3h}{2M\nu}}}{\sqrt{e^{\frac{h\nu}{kT}} - 1}} + s_0^{\frac{3}{2}} \right\}$$

an expression, which contains only α and s_0 as unknown constants. If we set

$$\begin{aligned} \sqrt{\frac{2\pi N \sqrt{3km}}{e^2 \alpha}} \cdot \frac{3h}{2M\nu\pi^2} &= A \\ s_0 \sqrt{\frac{2\pi N \sqrt{3km}}{e^2 \alpha}} &= B \end{aligned}$$

then W assumes the form given in the formula (70).

181 *F. A. Lindemann*, Phil. Mag. **29**, 127 (1915).
181a *F. Haber*, Berl. Akad. Ber. 1919, pp. 506 and 990.
182 *J. Stark*, Jahrb. d. Radioakt. u. Elektronik **9**, 188 (1912).
183 *G. Borelius*, Ann. d. Phys. **57**, 278 (1918).
184 *K. Herzfeld*, Ann. d. Phys. **41**, 27 (1913).

185 If we set $\frac{1}{2}mq^2 = E$, therefore $q = \sqrt{\frac{2E}{m}}$, the first of the two formulæ (72) follows from (67). If we further take into account that in *Drude's* Theory $E = \frac{3}{2}kT$, that is, that $k = \frac{2}{3}\frac{dE}{dT}$, then from (68) the second formula (72) follows.

186 *F. v. Hauer*, Ann. d. Phys. **51**, 189 (1916).
187 *W. Nernst*, Berl. Ber. 1911, p. 65.
188 *A. Eucken*, Berl. Ber. 1912, p. 141.
189 *K. Scheel* and *W. Heuse*, Ann. d. Phys. **40**, 473 (1913). Cf. also *L. Holborn*, *K. Scheel* and *F. Henning*, Wärmetabellen der physikal.-techn. Reichsanstalt (Vieweg 1919).
190 *A. Einstein* and *O. Stern*, Ann. d. Phys. **40**, 551 (1913).
191 The quantum formulæ (76) and (77) properly correspond to the *Planck* oscillator, that is, to a system of *one* degree of freedom, while here, in the case of rotation, we have to do with *two* degrees of freedom. But the energy of the *Planck* oscillator is composed of two equal parts, a kinetic and a potential part, while in the case of rotation only kinetic energy comes into question. This is often expressed thus: the *Planck* oscillator possesses one potential and one kinetic degree of freedom, while the rotating molecule possesses two kinetic degrees of freedom.
192 *P. Ehrenfest*, Verhandl. d. deutsch. physikal. Ges. **15**, 451 (1913).
193 According to note 48, the quantum canonical distribution function is

$$w_n = \frac{p_n e^{-\frac{E_n}{kT}}}{\sum_n p_n e^{-\frac{E_n}{kT}}}$$

and the mean energy is

$$\overline{E} = \frac{\sum\limits_{n} p_n E_n e^{-\frac{E_n}{kT}}}{\sum\limits_{n} p_n e^{-\frac{E_n}{kT}}}.$$

If we here set all p_n's $= 1$, and if for E_n we substitute the value $E_r^{(n)}$ from (80), there follows for the mean rotational energy of a molecule

$$\overline{E_r} = \frac{h^2}{8\pi^2 J} \cdot \frac{\sum\limits_{0}^{\infty} n^2 e^{-n^2\sigma}}{\sum\limits_{0}^{\infty} e^{-n^2\sigma}}, \quad \text{where } \sigma = \frac{h^2}{8\pi^2 J kT}$$

and for the heat of rotation of hydrogen we get the expression

$$C_r = 2N \frac{d\overline{E_r}}{dT}.$$

194 The turning impulse (moment of momentum) of a system, the mass-points of which possess the mass m_i, the velocities v_i, and the distances r_i from a fixed point (say the origin of co-ordinates), is a vector of the value

$$|\mathbf{U}| = \sum\limits_{i} m_i v_i r_i \sin (v_i r_i).$$

In the present case, the system consists only of the two atoms (mass M), which rotate around a circle of radius r with the constant velocity $v = r \cdot 2\pi\nu$.
Hence here

$$|\mathbf{U}| = p = 2Mr^2 \cdot 2\pi\nu = J \cdot 2\pi\nu,$$

where $J = 2Mr^2$ is the moment of inertia.

195 The impulse (or momentum) p_i corresponding to a generalised co-ordinate q_i is, according to note 48, defined by the relation $p_i = \dfrac{\partial \mathbf{L}}{\partial \dot{q}_i}$, where $\dot{q}_i = \dfrac{dq_i}{dt}$, and $\mathbf{L}$ is the kinetic energy of the system. Now here the angle of rotation ϕ is chosen as a generalised co-ordinate. But the kinetic energy of a body rotating about a fixed axis is known to be $= \frac{1}{2} \cdot$ (moment of inertia) $\times$ (angular velocity)^2, hence

$$\mathbf{L} = \frac{J}{2}\left(\frac{d\phi}{dt}\right)^2 = \frac{J}{2}\dot{\phi}^2.$$

Hence

$$p_\phi = \frac{\partial \mathbf{L}}{\partial \dot{\phi}} = J\dot{\phi} = J \cdot 2\pi\nu.$$

196 *F. Reiche*, Ann. d. Phys. **58**, 657 (1919).

197 The best curve was obtained by assigning the "weight" $2n$ to the nth quantum state of rotation. The *rotationless state* $(n = 0)$ thus receives the weight zero, i.e. it does not exist. This amounts to the same thing as the introduction of a zero-point rotation.

198 *E. Holm*, Ann. d. Phys. **42**, 1311 (1913).

199 *J. v. Weyssenhoff*, Ann. d. Phys. **51**, 285 (1916).

200 *M. Planck*, Ber. d. deutsch. physikal. Ges. **17**, 407 (1915).

201 *S. Rotszayn*, Ann. d. Phys. **57**, 81 (1918).

202 The curve is not drawn by *Planck*, but is discussed in the author's paper cited in note 196.

203 See likewise the author's paper quoted in note 196.

204 *P. S. Epstein*, Ber. d. deutsch. physikal. Ges. **18**, 398 (1916). Cf. also Phys. Zeitschr. **20**, 289 (1919).

205 *N. Bohr*, Phil. Mag. 1913, p. 857.

206 During "regular precession" the top turns uniformly about its axis of symmetry (axis of its figure), while at the same time this axis describes a cone of circular section about an axis fixed in space.

207 A compilation of the moments of inertia of the hydrogen molecule used by the various investigators is as follows :—

		$J \cdot 10^{41}$.
Einstein-Stern	. . . .	1·47
Ehrenfest	. . . .	0·69
Reiche	. . . .	$\left\{\begin{array}{c}2 \cdot 214\\2 \cdot 293\\2 \cdot 095\end{array}\right\}$ different curves.
Holm .	. . . .	1·36
Weyssenhoff	. . .	0·34
Rotszayn	. . . .	2·12
Epstein (Bohr's model)	. .	2·82

208 *N. Bjerrum*, Nernst Festschrift 1912, p. 90. *Bjerrum* did not, by the way, start from formula (79), but calculated with the values $\nu_n = \frac{nh}{2\pi^2 J}$, since, following a proposal of *H. A. Lorentz*, he set the rotational energy $E_r^{(n)}$ equal to $nh\nu_n$, in contrast to *Ehrenfest's* formulation (78), which rests on a sounder basis.

209 *S. P. Langley*, Annals of the Astrophysical Observatory of the Smithsonian Institution, Vol. I, p. 127, Plate XX (1900).

210 *F. Paschen*, Wiedem. Ann. **51**, 1; **52**, 209; **53**, 335 (1894).

211 *H. Rubens*, Berl. Ber. 1913, p. 513.

212 *H. Rubens* and *E. Aschkinass*, Wiedem. Ann. **64**, 584 (1898).

213 *H. Rubens* and *G. Hettner*, Berl. Ber. 1916, p. 167. See also *G. Hettner*, Ann. d. Phys. **55**, 476 (1918).

214 *W. Burmeister*, Ber. d. deutsch. physikal. Ges. **15**, 589 (1913).

215 *Eva v. Bahr*, Ber. d. deutsch. physikal. Ges. **15**, 710, 731, 1150 (1913).

216 Cf. *Lord Rayleigh*, Phil. Mag. **34**, 410 (1892). Let an HCl molecule, for example, be considered, which consists of a positively charged

hydrogen atom H+ and a negatively chlorine atom Cl− (see Fig. 13). Let its centre of gravity be S, and let a be the distance of the H+ atom from S. Let the line joining the two atoms be the axis of x', and let this axis turn in the positive direction about S at the rate of ν_r revolutions per second with respect to the fixed x-y-system. If, now, the two atoms vibrate relatively to one another with the frequency ν_0 and the amplitude A, then the x' co-ordinate of the H+ atoms may be represented thus

$$x' = a + A \sin (2\pi\nu_0 t).$$

If we project this vibration upon the fixed co-ordinate system, it follows that

$$\begin{cases} x = x' \cos (2\pi\nu_r t) = a \cos (2\pi\nu_r t) + A \sin (2\pi\nu_0 t) \cos (2\pi\nu_r t) \\ y = y' \sin (2\pi\nu_r t) = a \sin (2\pi\nu_r t) + A \sin (2\pi\nu_0 t) \sin (2\pi\nu_r t) \end{cases}$$

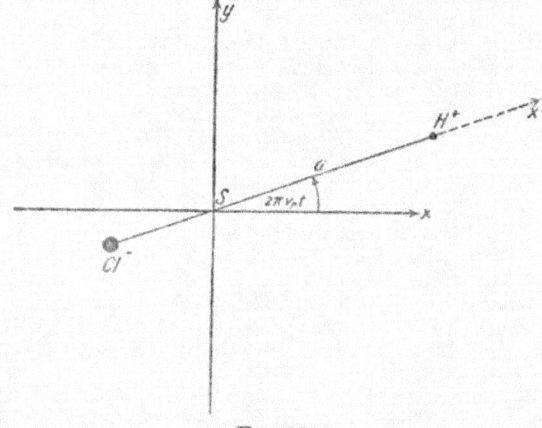

FIG. 13.

for which we may also write

$$\begin{cases} x = a \cos (2\pi\nu_r t) + \dfrac{A}{2} \sin 2\pi(\nu_0+\nu_r)t + \dfrac{A}{2} \sin 2\pi(\nu_0-\nu_r)t \\ y = a \sin (2\pi\nu_r t) - \dfrac{A}{2} \cos 2\pi(\nu_0+\nu_r)t + \dfrac{A}{2} \cos 2\pi(\nu_0-\nu_r)t. \end{cases}$$

From the point of view of the system at rest we have thus three oscillations :

(a) the left-circular oscillation

$$\left.\begin{array}{l} x = a \cos (2\pi\nu_r t) \\ y = a \sin (2\pi\nu_r t) \end{array}\right\} \text{ with the frequency } \nu_r$$

(b) the left-circular oscillation

$$\left.\begin{array}{l} x = \dfrac{A}{2} \sin 2\pi(\nu_0+\nu_r)t \\ y = -\dfrac{A}{2} \cos 2\pi(\nu_0+\nu_r)t \end{array}\right\} \text{ with the frequency } \nu_0+\nu_r$$

(c) the right-circular oscillation

$$x = \frac{A}{2} \sin 2\pi(\nu_0 - \nu_r)t \Bigg\}$$
$$y = \frac{A}{2} \cos 2\pi(\nu_0 - \nu_r)t \Bigg\} \text{ with the frequency } \nu_0 - \nu_r.$$

217 *E. S. Imes*, Astroph. Journ. **50**, 251 (1919).

218 *A. Eucken*, Ber. d. deutsch. phys. Ges. **15**, 1159 (1913). *Eucken* has here, on account of the asymmetrical form of the hydrogen molecule, assumed two different moments of inertia

$$J_1 = 0.96 \cdot 10^{-40}, \text{ and } J_2 = 2.21 \cdot 10^{-40}$$

and hence obtained two different series of numbers giving the revolutions ν_r per second, cf. the table given there. See also the table in *Rubens* and *Hettner*, loc. cit., note 213.

219 *M. Planck*, Ann. d. Phys. **52**, 491; **53**, 241 (1917).

220 *O. Sackur*, Ann. d. Phys. **36**, 958 (1911); **40**, 67 (1913).

221 *H. Tetrode*, Phys. Zeitschr. **14**, 212 (1913); Ann. d. Phys. **38**, 434 (1912).

222 *W. H. Keesom*, Phys. Zeitschr. **15**, 695 (1914).

223 *A. Sommerfeld*, Vorträge über die kinetische Theorie der Materie und der Elektrizität. Wolfskehl-Kongress in Göttingen 1913. (Teubner, Leipzig and Berlin 1914), p. 125.

224 *P. Scherrer*, Göttinger Nachr. 8 July, 1916.

225 *M. Planck*, Berl. Ber. 1916, p. 653.

226 *W. Nernst*, Die theoretischen und experimentellen Grundlagen des neuen Wärmesatzes. (W. Knapp, Halle 1918), pp. 154 *et seq.*

227 *O. Sackur*, Ber. d. deutsch. chem. Ges. **47**, 1318 (1914).

228 *O. Stern*, Phys. Zeitschr. **14**, 629 (1913); Zeitschr. f. Elektrochemie **25**, 66 (1919).

229 For what follows cf. the paper by *O. Stern* quoted in the last note. Further, *W. Nernst*, Die theoretischen und experimentellen Grundlagen des neuen Wärmesatzes. (W. Knapp, Halle 1918), Ch. XIII.

230 As regards this and the following chapter, the reader is referred for more exact details to the article of *P. S. Epstein* in the *Planck* number of "Naturwissenschaften" (1918, p. 230).

231 As the simplest *Thomson* atom, we are to imagine a sphere of radius a, filled with the unit charge e of positive electrification, of space-density ρ, in the middle of which an electron with the charge $-e$ rests. This structure is externally neutral. If we draw the electron out from the centre to a distance r (see Fig. 14) the external (shaded) hollow sphere exerts no force on the electron, according to the well-known laws of electrostatics. The inner solid sphere of radius r, on the other hand, acts on the electron just as if its total charge were concentrated at the centre.

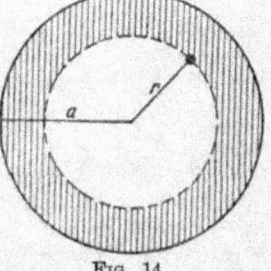

FIG. 14.

The force which draws the electron back into its position of rest is therefore,

$$F = \frac{\frac{4\pi}{3} r^3 \rho e}{r^2} = \frac{4\pi}{3} \rho e r = \frac{e^2}{a^3} r$$

that is, it is proportional to the distance of the electron from its position of equilibrium.

232 Cf. also P. *Drude*, Lehrbuch der Optik. 2. Aufl., Chs. V and VII (Hirzel 1906). There is an English edition of this work.

233 Cf. W. *Voigt*, Magneto- und Elektro-optik (Teubner 1908).

234 M. *Planck*, Ber. d. Berl. Akad. d. Wiss. 1902, p. 470; 1903, p. 480; 1904, p. 740; 1905, p. 382.

235 H. A. *Lorentz*, The Theory of Electrons, Chs. III, IV (Teubner 1909).

236 The electron oscillates, when bound quasi-elastically, according to the equation of motion $m \frac{d^2 x}{dt^2} = - fx$, if we restrict ourselves to *linear* oscillations. Here m is the mass of the electron, x is its distance from the position of rest, and f is a factor of proportionality. The solution of this differential equation is represented by the pure harmonic motion

$$x = A \cos (nt + \delta)$$

where the frequency is

$$n = \sqrt{\frac{f}{m}}.$$

The frequency n is therefore, as we see, independent of the amplitude and therefore of the energy of vibration.

237 The frequencies ν of those spectral lines of luminous hydrogen, which are included under the name "*Balmer* series," may be represented with great accuracy by the following formula given by *Balmer*.

$$\nu = N \left(\frac{1}{2^2} - \frac{1}{n^2} \right) \quad \text{where} \quad n = 3, 4, 5, 6 \ldots \infty.$$

N is here a constant, the so-called *Rydberg* number. If we set for the current number n the values $3, 4, 5 \ldots$ we get in succession the frequencies of the red line of hydrogen (H_a), the green line (H_β), and the blue line (H_γ) and so forth.

238 J. *Stark*, Ann. d. Phys. 43, 965 (1914); J. *Stark* and G. *Wendt*, ibid., 43, 983 (1914); J. *Stark* and H. *Kirschbaum*, ibid., 43, 991; 43, 1017 (1914); J. *Stark*, ibid., 48, 193, 210 (1915); J. *Stark*, O. *Hardtke* and G. *Liebert*, ibid., 56, 569 (1918); J. *Stark*, ibid., 56, 577 (1918); G. *Liebert*, ibid., 56, 589, 610 (1918); J. *Stark* and O. *Hardtke*, ibid. 58, 712 (1919); J. *Stark*, ibid., 58, 723 (1919).

239 Cf. H. A. *Lorentz*, The Theory of Electrons (Teubner, Leipzig and Berlin 1909), Ch. III.

240 H. *Geiger* and *Marsden*, Phil. Mag. April, 1913.

241 E. *Rutherford*, Phil. Mag. 21, 669 (1911).

242 According to C. G. *Darwin* [Phil. Mag. 27, 506 (1914)], the radius of the nucleus, taken as a sphere, is in the case of gold at the most $= 3 \cdot 10^{-12}$ cms., in the case of hydrogen at the most $= 2 \cdot 10^{-13}$ cms.

243 *A. van den Broek*, Phys. Zeitschr. **14**, 32 (1913).
244 Cf. note 247.
245 *N. Bohr*, Phil. Mag. **26**, 1, 476, 857 (1913).
246 *A. Einstein*, Phys. Zeitschr. **18**, 121 (1917).

247 The quite elementary calculation is as follows : let an electron of charge e and mass m rotate around a nucleus of charge $E = ez$ in a circular orbit : then z is the atomic number (for hydrogen, in particular, $z = 1$). If a is the radius of the circle, v the velocity, and ω the angular velocity (frequency of rotation) of the electron in the circular orbit, then the condition for equilibrium between the attraction of the nucleus and the centrifugal force is

$$\frac{eE}{a^2} = ma\omega^2 \quad \text{or} \quad ma^3\omega^2 = eE = e^2z.$$

According to *Bohr's* second hypothesis the moment of momentum $p(=mva=ma^2\omega)$ is a multiple of $\frac{h}{2\pi}$, hence

$$ma^2\omega = n\frac{h}{2\pi} \quad (n = 1, 2, 3 \ldots).$$

From these two equations for a and ω we get for the discrete radii of the permissible quantum orbits

$$a_n = \frac{n^2h^2}{4\pi^2e^2zm} \quad (n = 1, 2, 3 \ldots)$$

and the corresponding frequencies of rotation

$$\omega_n = \frac{8\pi^3e^4z^2m}{n^3h^3}.$$

The energy (kinetic + potential) is

$$W = \tfrac{1}{2}mv^2 + \left(-\frac{eE}{a}\right) = \tfrac{1}{2}ma^2\omega^2 - \frac{e^2z}{a} = -\frac{e^2z}{2a}$$

therefore the discrete quantum values of the energy are

$$W_n = -\frac{2\pi^2e^4z^2m}{h^2n^2}.$$

If, in this expression, we set

$$W = -\frac{e^2z}{2a} \text{ and } a = \left(\frac{e^2z}{m\omega^2}\right)^{\frac{1}{3}}$$

we recognise, that W is a function of ω, and hence of $\nu = \frac{\omega}{2\pi}$. The energy of the electron in the *Rutherford* model therefore depends, as stated in the text, on its frequency of rotation ν.

If the electron passes from the nth to the sth quantum path, then, according to *Bohr's* third hypothesis, a homogeneous spectral line is emitted of frequency

$$\nu = \frac{W_n - W_s}{h} = \frac{2\pi^2e^4mz^2}{h^3}\left(\frac{1}{s^2} - \frac{1}{n^2}\right) = Nz^2\left(\frac{1}{s^2} - \frac{1}{n^2}\right)$$

where

$$N = \frac{2\pi^2 e^4 m}{h^3}$$

248 Cf. note 237.

249 It is of historical interest to note that, before *Bohr*, *A. E. Haas* in 1910 (Sitzungsber. d. Wiener Akad. 10 March, 1910) succeeded in representing *Rydberg's* number in terms of the universal constants e, h, m; his result differed from that of *Bohr* only by a factor 8. He deduced his result as follows. Starting from *J. J. Thomson's* atomic model, which was generally accepted at that time, he calculated the maximum oscillation-frequency (no. of revolutions) ν_{max} of the electron in the simplest atom (hydrogen atom) for the case when this atom, provided with one energy-quantum, was circling just on the surface of the positive sphere. He obtained

$$\nu_{max} = \frac{4\pi^2 e^4 m}{h^3}.$$

This maximum frequency was next identified by *Haas* with the series limit ($n = \infty$) in *Balmer's* formula

$$\nu = N\left(\frac{1}{2^2} - \frac{1}{n^2}\right).$$

Then it follows that

$$N_{Haas} = \frac{16\pi^2 e^4 m}{h^3}.$$

which is a value 8 times greater than N_{Bohr}. *Haas* used this relation to calculate from the three quantities, the *Rydberg* number N, *Planck's* constant h, and the ratio $\frac{e}{m}$, all of which he assumed known, the charge e of the electron. In consequence of the factor 8 he obtained the value $e = 3\cdot18 \cdot 10^{-10}$, a value that is too small according to our present knowledge, but which agreed well with the measurements of *J. J. Thomson* and *H. A. Wilson*, which were available at that time.

250 *Th. Lyman*, Phil Mag. **29**, 284 (1915).

251 *F. Paschen*, Ann. d. Phys. **27**, 565 (1908).

252 *A. Fowler*, Month. Not. Roy. Astron. Soc. **73**, Dec. 1912.

253 *F. Paschen*, Ann. d. Phys. **27**, 565 (1908).

254 *E. C. Pickering*, Astroph. Journ. **4**, 369 (1896); **5**, 92 (1897).

255 *E. J. Evans*, Nature, **93**, 241 (1914).

256 *W. Kossel*, Ann. d. Phys. **49**, 229 (1916); Die Naturwissenschaften 7, 339, 360 (1919).

257 *L. Vegard*, Verhandl. d. deutsch. physikal. Ges. **19**, 344 (1917).

258 *A. Sommerfeld*, Atombau und Spektrallinien. (An English edition translated from the 3rd German edition (1922) is being prepared by Messrs. Methuen & Co., Ltd.)

259 *R. Ladenburg*, Die Naturwissenschaften **8**, 5 (1920).

260 *A. Sommerfeld*, Ann. d. Phys. **51**, 1 (1916).

261 Expressed in terms of polar co-ordinates the kinetic energy L assumes the well-known form :

$$L = \frac{m}{2}(\dot{r}^2 + r^2\dot{\phi}^2).$$

In it, m denotes the mass of the electron, the dots represent differentiation with respect to the time. The impulses p_r and p_ϕ are then defined as follows (see note 48) :

$$p_r = \frac{\partial L}{\partial \dot{r}} = m\dot{r} \; ; \qquad p_\phi = \frac{\partial L}{\partial \dot{\phi}} = mr^2\dot{\phi}.$$

262 Only when each impulse p_i depends solely on the corresponding q_i (or when it is a constant), and when, in addition, the limits of the phase-integral are independent of the q_i's, does the phase-integral work out to a constant. This is by no means the case for any arbitrary choice of the co-ordinate-system.

263 *P. S. Epstein*, Ann. d. Phys. **50**, 489 ; **51**, 168 (1916).

264 *K. Schwarzschild*, Sitzungsber. d. Berl. Akad. d. Wiss. 4. Map 1916.

265 *A. Einstein*, Verhandl. d. deutsch. physikal. Ges. **19**, 82 (1917).

266 *M. Planck*, Verhandl. d. deutsch. physikal. Ges. **17**, 407, 438 (1915); Ann. d. Phys. **50**, 385 (1916).

267 The semi-major axis of the ellipse, which is characterised by the values n and n', here has the value

$$a = \frac{h^2(n + n')^2}{4\pi^2 e^2 zm}.$$

The ratio of the axis is

$$\frac{b}{a} = \frac{n}{n + n'}.$$

We see that $n' = 0$ corresponds to the case of *Bohr's* circular orbits.

268 The energy of the electron moving in the *Kepler* ellipse (n, n') here has the value

$$W_{nn'} = -\frac{2\pi^2 e^4 z^2 m}{h^2(n + n')^2} = -\frac{Nhz^2}{(n + n')^2}.$$

The series formula (102) of the text then follows from *Bohr's* Law of Frequency

$$\nu = \frac{W_{nn'} - W_{ss'}}{h}.$$

269 If account is taken of the influence of relativity, the series formula for the spectra of the hydrogen type become to a first approximation

$$\nu = \nu_0 + \nu_1$$

where

$$\nu_0 = Nz^2\left[\frac{1}{(s + s')^2} - \frac{1}{(n + n')^2}\right]$$

$$\nu_1 = Nz^4\alpha^2\left[\frac{\frac{1}{4} + \frac{s'}{s}}{(s + s')^4} - \frac{\frac{1}{4} + \frac{n'}{n}}{(n + n')^4}\right].$$

In these expressions the symbols N and α have the following meaning:

$$N = \frac{2\pi^2 e^4 m_0}{h^3\left(1 + \frac{m_0}{M}\right)} \qquad \alpha = \frac{2\pi e^2}{hc}; \quad \alpha^2 \text{ is of the order } 5 \cdot 3 \cdot 10^{-5}$$

m_0 is the mass of the electron at vanishingly small velocities.

Hence whereas the first term ν_0 gives the old formula, which was obtained by neglecting the influence of relativity, the small additional term ν_1 represents the influence of relativity. As we observe, ν_1 does not only depend on the quantum sums $s + s'$ and $n + n'$, but also on the individual values s, s', n, n'. This member, ν_1, is thus responsible for the fine-structure.

270 If we apply the formula of the preceding note to H_α, we have to set $z = 1$, $s + s' = 2$, $n + n' = 3$. We then get

$$\nu = N\left[\frac{1}{2^2} - \frac{1}{3^2}\right] + N\alpha^2\left[\frac{\frac{1}{4} + \frac{s'}{s}}{2^4} - \frac{\frac{1}{4} + \frac{n'}{n}}{3^4}\right].$$

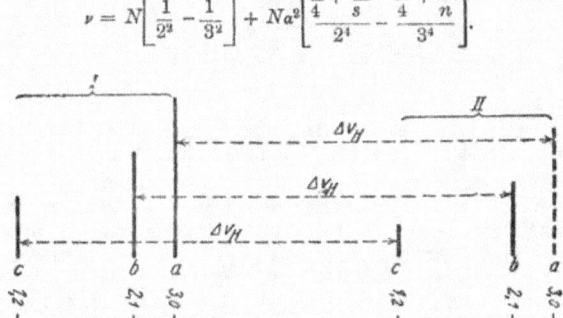

FIG. 15.

Corresponding to the possibilities of partition

$$s + s' = 2 = 2 + 0 \left.\begin{array}{l} \text{circle} \\ \text{ellipse} \end{array}\right\} 2 \text{ final orbits}$$
$$ = 1 + 1$$

and

$$n + n' = 3 = 3 + 0 \left.\begin{array}{l} \text{circle} \\ \text{ellipse} \\ \text{ellipse} \end{array}\right\} 3 \text{ initial orbits}$$
$$ = 2 + 1$$
$$ = 1 + 2$$

(for dynamical reasons the azimuthal quantum number n cannot under normal conditions assume a zero value), we should expect $2 \cdot 3 = 6$ possibilities of production and hence 6 components of the fine-structure of H_α. One of these components, however, namely, the one corresponding to the transition of the electron from the circle ($n = 3$, $n' = 0$) to the ellipse ($s = 1$, $s' = 1$) does not present itself under normal conditions, as follows from the " Principle of Selection " enunciated by *Rubinowicz* and *Sommerfeld* (see Chapter VI, §9). Hence 5 components of the fine-structure remain ; their position is exhibited in Fig. 15.

As we see, the 5 components arrange themselves into two main groups, containing 3 and 2 members, respectively. The "missing" line II_a is dotted in. The distance $\Delta\nu_H$ between I_a and II_a, I_b and II_b, I_c and II_c is called the "theoretical hydrogen doublet."

According to the above formula the frequency-number of the line I_a (3, 0→2, 0) is

$$\nu_{I_a} = N\left[\frac{1}{2^2} - \frac{1}{3^2}\right] + N\alpha^2\left[\frac{\frac{1}{4}}{2^4} - \frac{\frac{1}{3}}{3^4}\right].$$

The frequency-number of the line II_a (3, 0→1, 1) is

$$\nu_{II_a} = N\left[\frac{1}{2^2} - \frac{1}{3^2}\right] + N\alpha^2\left[\frac{\frac{1}{4}+1}{2^4} - \frac{\frac{1}{3}}{3^4}\right].$$

Thus

$$\Delta\nu_H = \nu_{II_a} - \nu_{I_a} = \frac{N\alpha^2}{2^4} = 1\cdot095 . 10^{10}$$

corresponding to $\Delta\lambda_H = 0\cdot157\overset{\circ}{A}$.

The hydrogen-doublet actually observed is measured from about the middle of I_a and I_b to the middle of II_b and II_c, owing to the absence of II_a. This leads to the value $0\cdot8\Delta\lambda_H$, that is, to $0\cdot126\overset{\circ}{A}$.

According to a principle of correspondence enunciated by *Bohr* (see Chapter VI, §9), as a result of which the azimuthal quantum number can only vary by ± 1, the components I_b and II_c are also absent.

271 *F. Paschen*, Ann. d. Phys. **50**, 901 (1916).

272 From formula (97) of the text we get for the two *Rydberg* constants for hydrogen and helium :

$$N_H = \frac{2\pi^2 m_0 e^4}{h^3\left(1+\dfrac{m_0}{M_H}\right)}$$

$$N_{He} = \frac{2\pi^2 m_0 e^4}{h^3\left(1+\dfrac{m_0}{M_{He}}\right)}$$

Moreover, according to note 269, we get the third formula giving the value of the constant for the fine-structure :

$$\alpha = \frac{2\pi e^2}{hc}.$$

From the first two relations, by using $M_{He} = 4M_H$, we get

$$\frac{m_0}{M_H} = \frac{\dfrac{e}{M_H}}{\dfrac{e}{m_0}} = \frac{N_{He} - N_H}{N_H - \frac{1}{4}N_{He}}$$

and hence

$$\frac{e}{m_0 c} = \frac{e}{M_{\mathrm{H}} c} \cdot \frac{N_{\mathrm{H}} - \frac{1}{4} N_{\mathrm{He}}}{N_{\mathrm{He}} - N_{\mathrm{H}}}.$$

The two *Rydberg* numbers N_{H} and N_{He} have been measured by *Paschen* with great accuracy :

$$N_{\mathrm{H}} = (109677 \cdot 691 \pm 0 \cdot 06) \, . \, c$$
$$N_{\mathrm{He}} = (109722 \cdot 144 \pm 0 \cdot 04) \, . \, c.$$

Moreover, $\dfrac{e}{M_{\mathrm{H}} \cdot c} = F$ is the electrochemical equivalent (*Faraday's* number), that is, the charge which, in electrolysis, accompanies one gramme-atom (i.e. $N = \dfrac{1}{M_{\mathrm{H}}}$ atoms). This number has the value

$$F = 9649 \cdot 4 \text{ electromagnetic units.}$$

If we insert the three values of N_{H}, N_{He} and $\dfrac{e}{M_{\mathrm{H}} \cdot c}$ in the relation above deduced, we get

$$\frac{e}{m_0 c} = 1 \cdot 7686 \, . \, 10^7 \text{ electromagnetic units,}$$

a value which agrees very well with those values of this quantity which were obtained by direct methods (deflection of the cathode- and β-rays in the electric and magnetic field). Let us now write

$$\frac{2 \pi^2 m_0 e^4}{h^3} = N_{\mathrm{H}} \left(1 + \frac{m_0}{M_{\mathrm{H}}} \right)$$

or, using the value of $\dfrac{m_0}{M_{\mathrm{H}}}$ given above,

$$\frac{2 \pi^2 m_0 e^4}{h^3} = \frac{3}{4} \frac{N_{\mathrm{H}} \cdot N_{\mathrm{He}}}{N_{\mathrm{H}} - \frac{1}{4} N_{\mathrm{He}}}.$$

The right-hand side of this equation is known. If we combine with it the value for $\dfrac{e}{m_0 c}$ just found, and also the value

$$a = \frac{2 \pi e^2}{h c} = 7 \cdot 290 \, . \, 10^{-3}$$

which follows from *Paschen's* measurements of the fine-structure in the case of helium, we have three equations in three unknowns e, m_0, h. From them we get

$$e = (4 \cdot 766 \pm 0 \cdot 088) \, . \, 10^{-10}$$
$$h = (6 \cdot 526 \pm 0 \cdot 200) \, . \, 10^{-27}.$$

According to *Sommerfeld* it is more advantageous to use *Millikan's* value for e. We then get

$$\begin{cases} e = (4 \cdot 774 \pm 0 \cdot 004) \, . \, 10^{-10} \\ h = (6 \cdot 545 \pm 0 \cdot 009) \, . \, 10^{-27} \\ a = (7 \cdot 295 \pm 0 \cdot 005) \, . \, 10^{-3}. \end{cases}$$

273 *K. Glitscher*, Ann. d. Phys. **52**, 608 (1917).

274 *A. Landé*, Phys. Zeitschr. **20**, 228 (1919); **21**, 114 (1920).

275 Cf. *A. Sommerfeld*, Atombau und Spektrallinien. Ch. IV, § 6.

276 *P. S. Epstein*, Ann. d. Phys. **50**, 489 (1916).

277 *P. Debye*, Göttinger Nachr. 3 June, 1916.

278 *A. Sommerfeld*, Phys. Zeitschr. **17**, 491 (1916). Cf. also Atombau und Spektrallinien. Ch. VI, § 5.

279 *F. Paschen* and *E. Back*, Ann. d. Phys. **39**, 897 (1912); **40**, 960 (1913).

280 *A. Rubinowicz*, Phys. Zeitschr. **19**, 441, 465 (1918).

281 *N. Bohr*, On the Quantum Theory of Line-spectra. Parts I and II. D. Kgl. Danske Vidensk. Selsk. Skrifter, Naturvidensk. og Mathem. Afd. 8, Raekke IV, 1. Kopenhagen 1918.

282 The number of revolutions of the electron per second in the sth quantum circle of *Bohr* is, in the case of hydrogen, according to note 247 :

$$\nu_s = \frac{w_s}{2\pi} = \frac{4\pi^2 e^4 m}{s^3 h^3}.$$

On the other hand, it follows from formula (93) of the text, if we take s considerably greater than 1 (high quantum numbers), and $n = s + 1$ (transition between neighbouring circles), that

$$\nu = \frac{2\pi^2 e^4 m}{h^3} \cdot \frac{(s + 1)^2 - s^2}{n^2 s^2}.$$

i.e.

$$\nu \text{ is of the order } \frac{2\pi^2 e^4 m}{h^2} \cdot \frac{2s}{s^4} = \frac{4\pi^2 e^4 m}{h^3 s^3} = \nu_s.$$

283 *P. S. Epstein*, Ann. d. Phys. **58**, 553 (1919).

284 *H. A. Kramers*, Intensities of Spectral Lines. D. Kgl. Danske Vidensk. Selsk. Skrifter, Naturvidensk. og Mathem. Afd. 8, Raekke III, 3. Kopenhagen 1919.

285 *A. Sommerfeld* and *W. Kossel*, Ber. d. deutsch. physikal. Ges. **21**, 240 (1919).

286 *J. Franck* and *G. Hertz*, Phys. Zeitschr. **20**, 132 (1919); in which references are also given. Cf. also *J. Franck* and *P. Knipping*, Phys. Zeitschr. **20**, 481 (1919); *J. Franck*, *P. Knipping* and *Thea Krüger*, Ber. d. deutsch. physikal. Ges. **21**, 728 (1919).

287 *J. Tate* and *Foote*, Phil. Mag. July, 1918.

288 References are given in the report by *J. Franck* and *G. Hertz*, mentioned in note 286.

289 *A. Einstein*, Phys. Zeitschr. **18**, 121 (1917).
Let us consider the two quantum states (1) and (2) of the atom, with the energies E_1 and E_2 ($E_2 > E_1$). The number of transitions $2 \to 1$ which take place in the time dt owing to radiation is then, according to *Einstein*, $N_2 A_{21} dt$, in which N_2 is the number of atoms in the state 2, and, therefore, according to note 48

$$N_2 = N w_2 = N C p_2 e^{-\frac{E_2}{kT}}.$$

N being the total number of atoms. A_{21} is a factor of proportionality.

The introduction of external monochromatic radiation of frequency ν and intensity $\mathbf{K}_\nu$ firstly brings about positive absorption, that is, transitions $1 \rightarrow 2$. The number of these in the time dt is, according to *Einstein*, $N_1 B_{12} \mathbf{K}_\nu$, in which B_{12} is a factor of proportionality, N_1 is the number of atoms in the state 1, and hence

$$N_1 = NCp_1 e^{-\frac{E_1}{kT}}.$$

Secondly, the external radiation also effects transitions $2 \rightarrow 1$ (negative absorption). The number of these that occur in the time dt $= N_2 B_{21} \mathbf{K}_\nu$, where B_{21} is a factor of proportionality. When the energy exchange is in equilibrium the number of transitions $2 \rightarrow 1$ must be equal to the number of transitions $1 \rightarrow 2$, hence

$$NCp_2 e^{-\frac{E_2}{kT}} A_{21} dt + NCp_2 e^{-\frac{E_2}{kT}} B_{21} \mathbf{K}_\nu dt = NCp_1 e^{-\frac{E_1}{kT}} B_{12} \mathbf{K}_\nu dt$$

i.e.

$$\mathbf{K}_\nu = \frac{\dfrac{A_{21}}{B_{21}}}{\dfrac{p_1 B_{12}}{p_2 B_{21}} \cdot e^{\frac{E_2 - E_1}{kT}} - 1}.$$

When the temperature increases indefinitely, $\mathbf{K}_\nu$ must also increase to infinitely great values ; from this it follows that

$$\frac{p_1 B_{12}}{p_2 B_{21}} = 1.$$

Finally, if we set $\dfrac{A_{21}}{B_{21}} = A$ for shortness, we get the relation given in the text :

$$\mathbf{K}_\nu = \frac{A}{e^{\frac{E_2 - E_1}{kT}} - 1}.$$

290 Cf. the résumé by *E. Wagner*, Phys. Zeitschr. **18**, 405, 432, 461, 488 (1917).

291 *G. Moseley*, Phil. Mag. **26**, 1024 (1913); **27**, 703 (1914).

292 *W. Kossel*, Verhandl. d. deutsch. physikal. Ges. **16**, 898, 953 (1914) ; **18**, 339 (1916).

293 *A. Sommerfeld*, Ann. d. Phys. **51**, 125 (1916); Phys. Zeitschr. **19**, 297 (1918). Cf. also Atombau und Spektrallinien. Ch. III, Ch. IV § 4, Ch. V § 5.

294 *L. Vegard*, Verhandl. d. deutsch. physikal. Ges. **1917**, pp. 328, 344 ; Phys. Zeitschr. **20**, 97, 121 (1919).

295 *P. Debye*, Phys. Zeitschr. **18**, 276 (1917).

296 *J. Kroó*, Phys. Zeitschr. **19**, 307 (1918).

297 *A. Smekal*, Wiener Ber. IIa **127**, 1229 (1918); **128**, 639 (1919); Verhandl. d. deutsch. physikal. Ges. **21**, 149 (1919). Cf. also *A. Smekal* and *F. Reiche*, Ann. d. Phys. **57**, 124 (1918).

298 *W. Kossel*, Zeitschr. f. Physik **1**, 119 (1920).

299 Since the *L*-ring consists of *several* electrons, we take the expression "elliptic motion" to mean the following type of motion : each electron independently describes an elliptic path about the nucleus, whereby the electrons are at each moment situated at the corners of a regular polygon which shares in the motion of the electrons, alternately contracting and expanding during this motion (" elliptical associates "), cf. *Sommerfeld.* Atombau and Spektrallinien.

300 *A. Smekal*, Wiener Ber. IIa **128**, 639 (1919).

301 *M. Born* and *A. Landé*, Berl. Akad. Ber. 1918, p. 1048 ; Verhandl, d. deutsch. physikal. Ges. **20**, 202, 210 (1918) ; *M. Born, ibid.*, **20**, 230 (1918) ; Ann. d. Phys. **61**, 87 (1920).

302 *A. Landé*, Verhandl. d. deutsch. physikal. Ges. **21**, 2, 644, 653 (1919) ; Zeitschr. f. Phys. **2**, 83 (1920). Cf. also *A. Lande* and *E. Madelung*, Zeitschr. f. Phys. **2**, 230 (1920).

303 *W. Kossel*, Ann. d. Phys. **49**, 229 (1916).

304 *P. Debye*, Münch. Akad. Ber. 9 Jan. 1915.

305 *P. Scherrer*, Die Rotationsdispersion des Wasserstoffs. Dissertation, Göttingen, 1916.

306 *G. Laski*, Phys. Zeitschr. **20**, 269, 550 (1919).

307 Cf., for example, *A. Sommerfeld*, Atombau und Spektrallinien, Ch. IV, § 6.

308 *Langmuir*, Journ. Amer. Chem. Soc. **34**, 860 (1912) ; Zeitschr. f. Electrochemie **23**, 217 (1917).

309 *Isnardi*, Zeitschr. f. Elektrochemie **21**, 405 (1915).

310 *J. Franck, P. Knipping* and *Thea Krüger*, Ber. d. deutsch. physikal. Ges. **21**, 728 (1919).

310a *Planck* has made an attempt to alter *Bohr's* model in such a way that the right heat of dissociation results. See *M. Planck*, Berl. Akad. Ber. 1919, p. 914. Cf. also *H. Kallmann*, Dissertation, Berlin 1920.

311 *W. Lenz*, Ber. d. deutsch. physikal. Ges. **21**, 632 (1919).

312 *A. Sommerfeld*, Ann. d. Phys. **53**, 497 (1917).

313 *F. Pauer*, Ann. d. Phys. **56**, 261 (1918).

314 *G. Laski*, see note 314.

315 *M. Pier*, Zeitschr. f. Elektrochemie **16**, 897 (1910).

316 *K. Schwarzschild*, Berl. Akad. Ber. 1916, p. 548.

317 *H. Deslandres*, Compt. Rend. **138**, 317 (1904).

318 *T. Heurlinger*, Phys. Zeitschr. **20**, 188 (1919) ; Zeitschr. f. Physik **1**, 82 (1920).

319 *W. Lenz*, see note 311.

320 A different view is upheld by *J. Burgers* (Versl. K. Ak. van Wet. Amsterdam **26**, 115, 1917), in which, also, jumping electrons produce the middle line in the infra-red of band. In contrast with *Schwarzschild* and *Lenz, Burger* assumes that the motion of the electrons is influenced by the rotation of the molecule. The energy of the system is then not composed additively of the energy of the electrons and the rotational energy of the molecule, but a third term has to be added, which is due to the Coriolis force of the rotating system.

[321] T. *Heurlinger*, see note 318.

[322] F. *Reiche*, Zeitschr. f. Physik 1, Heft 4, 283 (1920).

[323] E. S. *Imes*, Astrophys. Journ. 50, 251 (1919).

[324] A. *Kratzer*, Dissertation, München 1920.

[325] In the case of the gases investigated by *Imes*, namely HCl, HBr, and HF, the following moments of inertia were found:

$$J_{HCl} = 2 \cdot 64 \cdot 10^{-40}; \qquad J_{HBr} = 3 \cdot 27 \cdot 10^{-40}; \qquad J_{HF} = 1 \cdot 37 \cdot 10^{-40}.$$

INDEX

181

PRINTED IN GREAT BRITAIN AT THE UNIVERSITY PRESS, ABERDEEN

www.ingramcontent.com/pod-product-compliance
Lightning Source LLC
Chambersburg PA
CBHW071427170526
45165CB00001B/426

* 9 7 8 1 4 5 2 8 8 8 5 0 7 *